Samira ZAMICHE

Medical parasitology

Samira ZAMICHE

Medical parasitology

ScienciaScripts

Imprint

Cover image: www.ingimage.com

This book is a translation from the original published under ISBN 978-620-6-72613-5.

Publisher:
Sciencia Scripts
is a trademark of
Dodo Books Indian Ocean Ltd. and OmniScriptum S.R.L publishing group

120 High Road, East Finchley, London, N2 9ED, United Kingdom
Str. Armeneasca 28/1, office 1, Chisinau MD-2012, Republic of Moldova, Europe
Managing Directors: Ieva Konstantinova, Victoria Ursu
info@omniscriptum.com

Printed at: see last page
ISBN: 978-620-8-62457-6

FOREWORD

Medical parasitology is an essential discipline for understanding infectious diseases caused by parasites. These diseases, which affect millions of people worldwide, represent a major public health challenge, particularly in developing countries. With the increase in international travel and climate change, these parasitic infections now affect a wider population, underlining the importance of preventing and combating these diseases.

The aim of this book, Medical Parasitology, is to provide a comprehensive and rigorous overview of the main parasites affecting humans and the diseases they cause. Aimed at students, health professionals, researchers and anyone with an interest in medical biology, this book presents not only the biological and epidemiological aspects of parasites, but also strategies for diagnosis, treatment and prevention.

Browsing through these pages, you will discover the transmission mechanisms of parasites, their complex biological cycles, and the different therapeutic approaches available. Each chapter is structured to provide an in-depth understanding of parasites, from protozoa to helminths and arthropod vectors. We have also included sections on current challenges, such as resistance to antiparasitic drugs and new diagnostic technologies.

This book has been written with a view to providing clear, precise and up-to-date information, based on the most recent research in the field. We hope that it will become a useful reference for students of parasitology, as well as a valuable working tool for healthcare professionals involved in the diagnosis and treatment of parasitosis.

This book is the fruit of several years' work, teaching and research. We would like to express our gratitude to all those who have contributed to the production of this book, whether colleagues, students or practitioners, whose suggestions and feedback have enriched the content. We hope that this publication will make a significant contribution to the study of parasites and the diseases they cause, and that it will encourage future research in this crucial area.

I- GENERAL INFORMATION ON PARASITISM

1. parasitism

Parasitism involves a close association between two different organisms. One of them, the host, provides food and shelter for the second, the parasite. The parasite may or may not be harmful to the host. A parasite (from the Greek para= next to; sitos= food) is a living being that evolves at the expense of another living being without destroying it completely. In many cases, the parasite exerts a pathogenic action on the host, causing damage at certain stages of its evolution or when there are too many of it.

Parasitism must be distinguished from other forms association such as :
- **commensalism:** association between two different individuals, benefiting one, the commensal, and causing no harm to the host (e.g. Entamoeba coli living in the human intestine).

- **symbiosis: an** association between two living beings for mutual benefit (e.g. ciliated protozoa in the rumen of ruminants that digest plant cellulose and transform it into glucose, a simple sugar that can be assimilated by the host organism).
- **predation:** association between two individuals, one of which is the predator (e.g. lion, panther, wolf, etc.) kills another, the prey (e.g. antelope, buffalo, wildebeest, hare, etc.) to ensure its subsistence.

1.1.Classification

Parasites are classified into four main groups: micromycetes, helminths, arthropods and protozoa.

-**Micromycetes** belong to the Fungi kingdom. Some species are parasites, causing disease in plants, animals and other fungi.

- **Helminths or worms** are multicellular invertebrate organisms belonging to the animal kingdom, which may or may not be parasitic. They are divided into two main groups: flatworms and roundworms.

- **Arthropods** belong to the animal kingdom and are ectoparasites responsible for many diseases, as well as being vectors of many human and animal diseases.

- **Protozoa** are protists, single-celled, heterotrophic, motile eukaryotic organisms. Most protozoa are free-living or live in symbiosis. However, some species are pathogenic parasites responsible for disease.

2. Different types of parasitism

- **Accidental parasitism** Certain free-living animals, such as myriapods (centipedes) and insect larvae, are capable of passing into a host and causing trouble. These are known as accidental parasites.

- **Facultative parasitism** Facultative parasites are animal or plant living beings which normally live in decomposing organic matter and which, in certain environmental conditions (wounds, contusions, etc.) can pass from a free life to a parasitic life (e.g. the larvae of certain flies which usually develop on corpses).

- **Obligate parasitism** These are individuals that are closely dependent on their host for their subsistence and are therefore unable to lead a free life outside the host (e.g. roundworms, tapeworms, fleas, lice, etc.). Compulsory parasites include :

- **Temporary parasites** (e.g. mosquitoes, horseflies, ticks): these are parasites that leave their hosts after having eaten.
- **Permanent parasites:** some parasites never leave their host (e.g. lice, mange mites) or are only present in the environment eggs (roundworms, tapeworms, pinworms, etc.).
- **Periodic parasites:** many living things are parasites for part of their lives:

- **or in the adult stage** for fleas, mosquitoes, horseflies, etc.
- **or at the larval stage** for Oestrus ovis larvae in sheep, or Hypoderma bovis in cattle...

3. The relationship between parasites and their host

3.1 Parasite specificity: this is the bond of more or less strict fidelity which unites a parasite to its host or hosts:
- **Stenoxenous parasites** (sténos: narrow; xénos: host): are those that are very closely adapted to a single host and cannot live at the expense of other hosts (e.g. Pou, Plasmodium, Taenia);
- **Oligoxenous parasites** (oligos: few; xenos: host): some parasites have a less narrow parasite specificity and can be found in hosts belonging to neighbouring species or genera (e.g. Diphyllobothrium latum: tapeworm parasite of the small intestine of humans, dogs and cats).
- **Euryxenic parasites** (euryos: broad; xenos: host): on the other hand, are only loosely specific. They can move from one host to another with varying degrees of ease (e.g. haematophagous insects that carry diseases common to humans and animals).

3.2.Habitat or location of parasites

Depending on the location, a distinction is made between :
- **ectoparasites or external parasites** (e.g. fleas, ticks, lice...) living on the surface of the skin
- **endoparasites or internal parasites** (e.g. tapeworms, roundworms, hydatid cysts, digestive and respiratory strongyles, liver flukes, etc.) living in the various tissues and deep cavities of the body.

3.3. Nutrition

Parasites feed in different ways:
- Parasites with a digestive tract (e.g. nematodes or roundworms) feed by capturing the products digestion (intestinal chyme). Nutrients are assimilated by enzymes secreted by the parasites.

- Some parasites without a digestive tract (flatworms: cestodes and trematodes) and parasitic fungi feed by osmosis.

eat wide variety of foods: some are haematophagous (e.g. ticks, fleas, mosquitoes, flukes), others are histophagous e.g. mange mites, liver flukes) and feed on tissues or desquamated cells. Others are chymivorous (e.groundworms).

3.4. Parasite adaptation

Parasite adaptation most often results in morphological changes, biological and physiological.

a/ Morphological adaptations

The transition of certain organisms to parasitic life is reflected in :

-atrophy or reduction of certain organs (e.g. sense organs limited to small papillae in the head, neck or mouth of nematodes).

- by the disappearance of certain locomotor appendages and the circulatory system.
- by the regression of digestive and respiratory systems.

Parasitism, on the other hand, is responsible for the appearance of new morphological features such as attachment organs (suckers, tapeworm hooks) that enable the parasite to adhere its host. Other parasites, on the other hand, have not been modified by parasitism (e.g. flagellate and ciliate protozoa).

b. Physiological adaptations

Some parasites have been able to adapt to environments little or no oxygen (anaerobiosis) and do not suffer. Other parasites have succeeded in developing immunity to host enzymes (e.g. L1 stage larva of Hypoderma bovis migrating in bovine muscle masses). Parasites secrete substances that neutralise action of host enzymes. 6.3)

c. Biological adaptations

The biological adaptation of parasites to a given host is linked to the parasite's subsistence needs and to a set of tropisms. According to Brumpt (1949), tropisms are defined as "regular, precise, inevitable phenomena appearing a series excitations of living beings from the external environment".

There are different types of tropism depending on the species of parasite involved, its stage of development, its sex, its physiological state and its location. A distinction is made between :

-histotropism: the ability of the parasite to penetrate the host's tissues (e.g.: the ability of the arasite to penetrate the host's tissues):

Trichina larva);

- haematotropism: predilection of parasites for blood (e.g. Babesia bovis; Trypanosoma berberum).

- dermotropism: parasites that live in the skin (e.g. mites that cause scabies).

3.5. Action of parasites on the host.

The parasite introduces 4 types action that can be exerted on the host.

1- Spoliating action: This is more or less important. All parasites grow at the expense of the organism from which they steal part of the assimilable substances (vitamins, blood, etc.) which were intended for it. In some cases it is insignificant, in others it is very important. Many parasites feed on blood and have an obvious spoliating role: haematophagous arthropods (e.g. fleas, mosquitoes, ticks, leeches, horseflies, etc.). Some internal parasites (e.g. hookworms) draw blood from their host's small intestine and waste a lot of it. Other parasites exert a selective spoliating action on the intestinal chyme and consume large quantities of carbohydrates (e.g. roundworms). Others feed on the host's tissues and take tissue fragments from the parasitized host's body (e.g. young liver flukes feeding on liver tissue).

2- Toxic action: This is due to the emission (excretion or secretion) of toxic substances by the parasite, which cause reactions in the organism. Haematophagous arthropods inoculate the bite wound of a host with toxic substances contained in their saliva, sometimes causing strong inflammatory and nervous reactions in the host.

For example, the protozoan parasites Babesia spp, Trypanosoma berberum and Leishmania infantum can be highly toxic, causing necrotic lesions in various organs and nervous disorders in parasitised individuals.

Other parasites produce toxic products either at the adult stage (e.g. roundworm) or at the larval stage (e.g. L1 larva of Hypoderma bovis) and store them in their general cavity. Their massive destruction following treatment allows release of large quantities of toxins that cause serious disorders in animals.

3- Infectious action: The larvae of certain worms perforate the skin or mucous membranes and undertake complex migrations within the body, leading to serious infections. Some parasites coexist with certain microbes, so their introduction into the host organism allows the microbe to penetrate. Similarly, certain helminths (worms) can promote infection by increasing the general susceptibility of parasitised animals (e.g. during their migration, young flukes can carry spores of anaerobic germs (Clostridia) from the intestine to the liver, where they responsible for necrotising hepatitis.

4- Mechanical action: During their development, some parasites have a mechanical compressive action on organs (e.g. hydatid cysts). Other parasites, when numerous, can cause intestinal or vesicular obstructions (e.g. obliteration of the bile duct by roundworm larvae).
5- Traumatic action: Haematophagous arthropods (ticks, fleas, mosquitoes, etc.) use their mouthparts to exert a traumatic action by perforating the skin when bitten. Certain helminth larvae (worms) break through the integument during the penetration phase (e.g. L3 stage larva of Ankylostoma caninum) or through the intestinal wall (e.g. larva of Toxocara canis), and migrate to various parts of body.
6- Transmission of pathogenic germs: Biting arthropods inoculate various pathogenic germs (parasites, bacteria, viruses, etc.) during their blood meal:

For example, sandflies, which are haematophagous insects, infect dogs with Leishmania canis.

3.6. Host reaction

The presence of parasites triggers two types of defence reaction in the host: cellular and humoral.

3.4.1. Cellular reactions

a) **Phagocytosis:** this is the first barrier to parasitic infestation. This means of defence can sometimes favour the evolution of certain parasites whose habitat is the protoplasm of macrophages (e.g. multiplication of Leishmania infantum in the cells of the S.R.H.). Other cellular reactions may occur if phagocytosis is insufficient to neutralise the parasites.

b) Inflammatory reactions: these are characterised by conjunctivo-vascular neoformations which are more or less rich in various leucocytes and in the centre of which the parasite is sometimes encysted (e.g. fibro conjunctiva shell of the hydatid cyst). In addition to the connective tissue, the nodular reactions observed around the larvae are made up of eosinophilic cells and giant cells.

c) **Eosinophilia:** blood eosinophilia is also a means of defending the body against various toxins, but it is not specific to parasitic diseases. It is also present in allergy .

In parasitology, **eosinophilia** is particularly marked in helminthosis (diseases caused by parasitic worms).

3.4.2 Humoral reactions

The presence of parasites and their toxins induces the body to synthesise various antibodies

4. Cycle

To guarantee the continuity of its own species, the parasite must undergo transformations and follow sometimes highly complex evolutionary paths, forming what is known as the parasite cycle. Given the complexity of some of these cycles, it is surprising to see the persistence of species whose chances of survival seem uncertain. Parasites compensate for this handicap by :

• extraordinary fertilisation: T.saginata 150 million eggs a year:
•polyembryony: intense multiplication at the larval stage (hydatid, budding larval forms of flukes);

• remarkable resistance in the external environment (roundworm eggs survive for years):
•a very long lifespan in the host: tens of years for Plasmodium, Anguillaria and Filaria;

• metabolic and immunological adaptation (mimicry): anti-enzymes (Taenia), modification of anti-gene structure (schistosomes).

4.1.Different types of evolutionary cycles

The different cycles can be summed up as the number of hosts required a parasite to pass on to the next generation. It is always in the larval stage that a parasite arrives at a host and settles on or in the site (organ) that suits it, to become an adult and reproduce. This host is always the definitive host.

The two main types of biological cycles are :
• The direct cycle: monoxene or holoxene cycle
• Indirect cycles: heterogenous cycles

a. The direct cycle or monoxene or holoxene cycle

The parasite completes its entire cycle in the same host or in part in the external environment. Parasites are said to be monoxenous (without passing through the external environment: lice, scabies mites, etc.; with passing through the external environment: roundworms, digestive strongyles, etc.).

b. **The indirect or heteroxene cycle**

the evolutionary cycle of the parasite takes place in several hosts: the parasite lives as adult in a host called the definitive host, and evolves as a larva in one or more hosts called intermediate hosts. Parasites are referred to as heteroxenous parasites (e.g. Echinococcus granulosus: adult stage of the parasite in dogs; larval stage in ruminants and humans). Most often, the definitive host (H.D.) harbours the sexed forms or adult stages and the host intermediate (H.I.) houses the asexual forms or larval stages.

• The final host (HD)

By definition, it is the living creature that harbours the adult form of the parasite. This will not always man in medial pathology (hydatidosis, cysticercosis).However, in malaria, although the sexed forms develop in From the practical epidemiological (not parasitological) point of view, humans considered to be HD for Plasmodium.

• **The intermediate host (HI)**

In the epidemiological chain, it is situated between the parasite reservoir and the receptive host.To speak of an intermediate host in the true sense of the word, the parasite must require passage through this host to ensure its maturation or true multiplication, and often both.
Infesting forms only appear after the biologically essential passage through this intermediate host.
These intermediate hosts can be differentiated into :
H.I. Passive: They are content host the infesting form or its anterior stages other than the adult form, without approaching the parasite reservoir to contract the parasite, or the receptive host to transmit the infesting forms. E.g. molluscs for flukes or schistosomes; cyclops for the Medina filaria; beef or pork for tapeworms.
Active H.I.: Not only do they host the developing parasite, but they also move towards the parasite reservoir to contract the parasitism and towards the receptive host to transmit or inoculate it.
These H.I. are vectors (tsetse flies for sleeping sickness, mosquitoes for Bancroft's threadworm, horseflies for Loa loa threadworm).

Vectors

They are the transmitters of parasites.

A distinction can be made between biological vectors which are essential to the parasite's life cycle and which ensure maturation or multiplication. These are the H
.I. active (haematophagous arthropods).
Purely mechanical vectors which have a simple phonetic function (transport) which is not essential for the parasite's life cycle (e.g. flies: transport amoeba cysts - support cry of moat metacercariae).

• **The parasite reservoir (R.P.)**

This is the biotope in which the parasite survives. Generally speaking, outside a radial environment (soil: telluric RP) it will be a definitive host that houses the parasite.

4.2. Epidemiology and modes infestation :

host can be infested in different ways, depending on the parasite's cycle:

- The free-living forms generally infest orally, transdermally or sexually.
- Some forms are transmitted by an intermediate host, either orally (ingestion of the HI or ingestion of parasitised meat), or insect bites (deposition of saliva, faeces or secretions).
- By blood transfusion
- By grafting a parasitized organ

5. Diagnosis

Diagnosis of parasitic infections is based on biological analyses. These involve detecting the parasite in a biological sample, or observing the body's reaction to the parasite. The parasite can be detected directly in the sample by :

- Macroscopic observation (in the case of parasites visible to the naked eye) ;
- Microscopic observation
- Immunological analysis
- Histology
- Molecular biology techniques;

It can also be made indirectly by culturing the parasite on a medium, or by inoculating an animal. Biological diagnosis is based on the biological changes caused by non-specific host responses (such as anaemia, hypereosinophilia, etc.) or by specific reactions of the immune system (detection of specific antibodies).

6. Prophylaxis

The aim of prophylactic measures is to combat parasitic diseases. The aim of prophylaxis is to break the evolutionary cycle of parasites. To this, two types of measures are applied: offensive measures or general prophylaxis and defensive measures or individual prophylaxis.

6.1. General prophylaxis: The aim is to destroy parasites wherever they are found: in the host's body, in the external environment and in intermediate hosts where they exist.

6.1.1. In the host: medical treatment using specific products (anthelmintics, insecticides, antifungals, etc.) and sanitary treatment using physical agents (cold, heat).

6.1.2. In the external environment :

- draining ponds to eliminate breeding grounds for various parasites.
- spreading insecticides and molluscicides.
- destruction of faeces from animals parasitised by fire or the biothermal method.

6.1.3. Action on intermediate hosts and vectors

- Vector control is essentially based on the use of chemical means (insecticides, acaricides, etc.).

6.2. Individual prophylaxis This consists of preventing infestation by various means: vaccination and hygiene measures.

II. PROTOZOA PARASITES

- Kingdom: Protista
- Phylum of Sarcomastigophora (Rhizoflagellates)
- Subphylum Mastigophora (Flagellates)
- Class of Zoomastigophora (or Zooflagellates)
- Order Kinetoplastida
- Suborder Trypanosomatina
- Trypanosomatidae family

Trypanosomes are protozoan parasites, unicellular eukaryotes, belonging to the order Kinetoplastida.

The order Kinetoplastidae comprises a group of protozoa with a single giant mitochondrion inside which the mitochondrial DNA is compacted, forming a dense structure called a kinetoplast. The suborder Trypanosomatina comprises a single family Trypanosomatidae with a single flagellum (Levine et al. 1980). Trypanosomatidae infect all classes of vertebrates and certain invertebrates such as plants.

Of all the Trypanosomatidae, two genera are known to be major pathogens: Leishmania spp and Trypanosoma spp. They are responsible for diseases in humans and other animals worldwide, causing major societal and economic losses

Trypanosomid parasites of humans belong either to the genus **Trypanosoma** or to the genus **Leishmania;** they are fairly complex parasites often requiring the presence of several hosts in the course of their development cycles, during which they undergo a succession of morphological stages.

II.1.Species: Trypanosoma brucei, responsible for human African trypanosomiasis (HAT) or sleeping sickness

Human African trypanosomiasis (HAT), also known as sleeping sickness, is caused by the parasite T. brucei, transmitted to the final host by its vector the tsetse fly. Despite rare reports self-healing, sleeping sickness is fatal without treatment.

HAT is caused by two subspecies of T. brucei

T. b. rhodesiense

T. b. gambiense.

• **Morphology :**

T. brucei is a unicellular flagellate parasite measuring approximately 20-25 µm long and 3-5 µm wide for the procyclic form and 15 µm long and 2 µm wide for the blood form. The parasite's shape is maintained by a cytoskeleton of sub-pellular microtubules (Fig.1). It has a flagellum, which gives it motility and mobility, emerging from the posterior part of the parasite at the level of the basal bodies, close to the flagellar pouch. The flagellum is attached to the whole of the cell body, up to the anterior part where its end is free. The mitochondrial genome is compacted into a dense structure called the **kinetoplast**, a large structure that can be revealed by specific DNA dyes.

Trypanosomids are parasitic flagellates with a single forward-facing flagellum, often connected to the body by **a waving membrane** (Fig.1). They are characterised by the presence of **a kinetoplast** (a differentiated part of the mitochondrion that contains ¼ of the

cell's DNA).
giant mitochondria of Trypanosomidae). The kinetoplast is always linked to the blepharoplast of the flagellum. They have a single flagellum running along the cell surface. This so-called recurrent flagellum appears as an 'undulating membrane'.

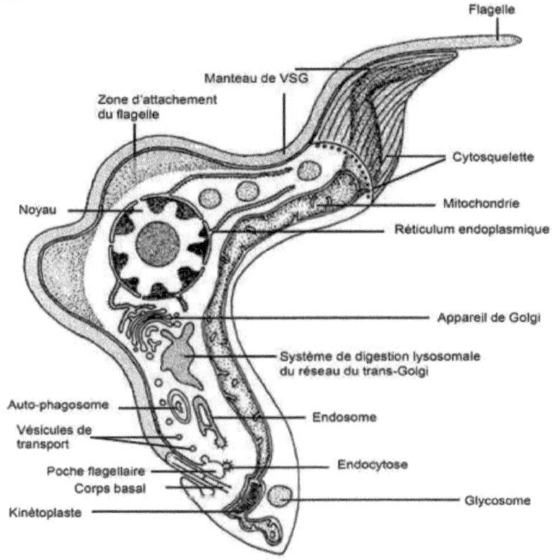

Figure 1: Schematic diagram of cellular architecture of the bloodstream form of Trypanosoma brucei

Adapted from (Vickerman, Myler, and Stuart 1993)

Trypomastigote form: This is an elongated form with a central nucleus, a posterior kinetoplast, an undulating membrane running the length of the body and a free flagellum from anterior end (Fig 2a).

The cell body measures 15 to 20 µm in length (the size of 3 red blood cells), is mobile in vertebrate blood and multiplies by binary division.

Note: there is a stockier form known as "infectious metacyclic", which is the evolutionary end of HI and ensures contamination of the HD.

Epimastigote form: This is an elongated form with a central nucleus, a kinetoplast close to the nucleus, an undulating membrane running along the body from the nucleus and a free flagellum from the anterior end (Fig2b). The cell body is 15 to 20 µm long and is found in the intermediate host.

Pro-mastigote form: it is elongated and has a central nucleus and a very anterior kinetoplast located at the base of the flagellum which measures 10 to 15 µm. The cell body measures 8-24 x 4- 5 µm and is highly mobile, moving forward with the flagellum. This form multiplies by scissiparity and is found in the digestive tract of intermediate hosts.

Amastigote form: This form differs greatly from the three previous ones by its very small size (2 to 4 microns) (Fig2c), by its rounded shape and by the absence of the free flagellar whip (the flagellum is reduced to a short internal portion). This form is therefore immobile and adapted to intracellular life. By contrast, the other three forms are mobile and adapted to life in the environment.

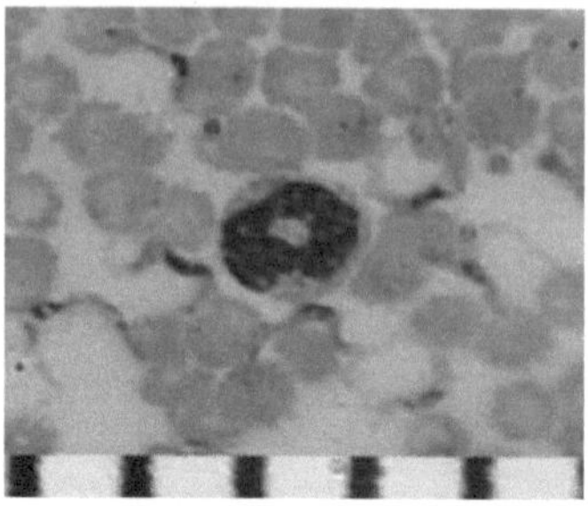

Fig 2a: Trypomastygote form

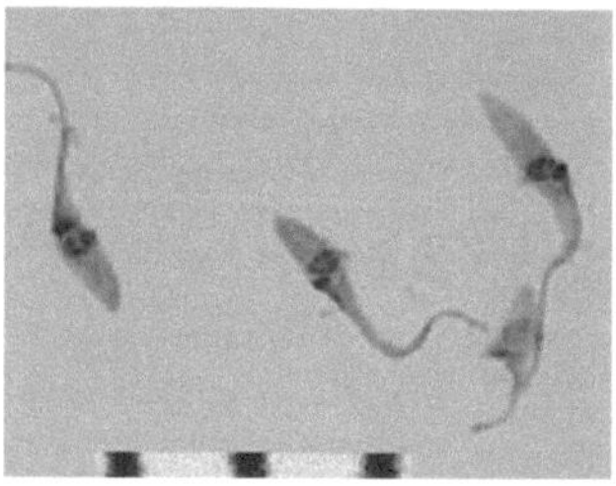

Fig 2b: Epimastigote form

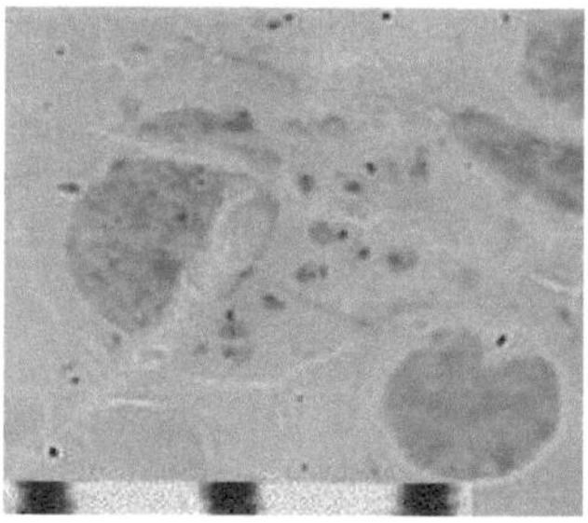

Fig 2c: Amastigote form

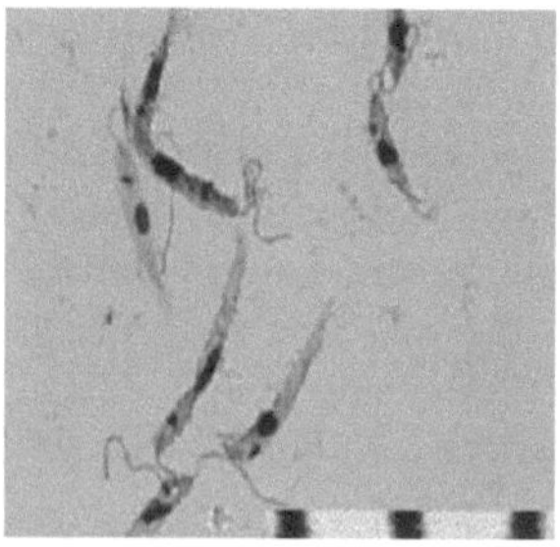

Fig2d: Promastigote form

The trypomastigote and amastigote forms are found in the vertebrate final host, while the promastigote and epimastigote forms are found in the insect intermediate host.

- **Evolutionary cycle :**

T. brucei has a complex life cycle, transmitted to the mammalian host by its vector, the tsetse fly. The development cycle alternates between **2 hosts** (tenene cycle):

- A vertebrate (mammals=HD) where they are found in the blood.
- The other invertebrate (tsetse fly=HI) where they are located in the digestive tract. Within these two hosts, it will undergo changes in morphology (Fig3).

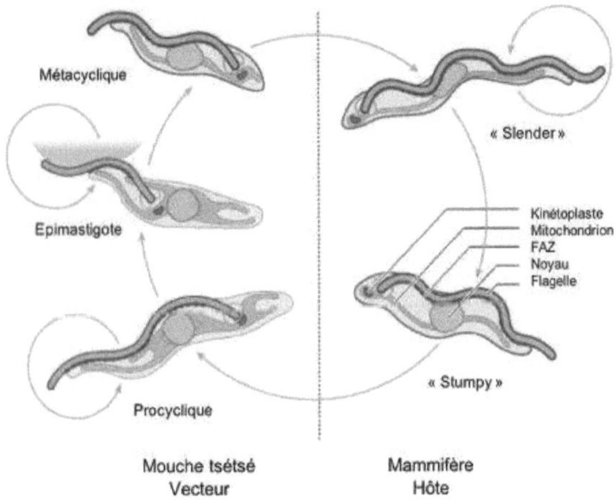

Figure 3: Life cycle of Trypanosoma brucei

Definitive host (HD): male for T. brucei.

When an infected tsetse fly bites a healthy host, it releases **metacyclic trypanosomes**, present in the fly's saliva, into the bloodstream.

The metacyclic forms differentiate into a long replicating trypomastigote blood form known as "slender", which multiplies in the blood. It can also invade other organs, particularly the brain and central nervous system (CNS).

The "slenderform can also differentiate in the blood or other organs and tissues into a short non-replicating trypomastigote form known as "stumpy". The "stumpy" form is pre-adapted to the fly's environment, with the appearance of a coat of procyclin and the reactivation of the mitochondrial respiratory chain (Fig.3).

Intermediate host (HI): The intermediate host is Glossina sp, it is commonly called the Tsetse fly (Fig4) which is a dipteran insect **whose males and females are haematophagous.** Glossina palpalis transmitted T. gambiense

When a tsetse fly bites a host infected with T. brucei, it ingests trypanosomes its blood meal.

Inside the fly's stomach, a few hours after ingestion, the trypanosomes differentiate into **a**

procyclic form. The procyclic form is replicative, multiplying inside fly's midgut and then migrating towards the salivary glands.

During migration, the procyclic form differentiates into the **epimastigote form**. The epimastigote form is characterised by asymmetric division, with the formation of a long epimastigote and a short epismastigote. The long form is replicative while the short form is non-replicative. The trypanosomes migrate to the salivary glands, at which stage the short epimastigote forms attach to the salivary gland epithelium by their flagellum and elongate, eventually differentiating **into a metacyclic form** which is released into the lumen. **The metacyclic form is the infectious form** pre-adapted to the host environment, and in particular has **a coat of variable surface glycoprotein (VSG).** The duration of development inside the tsetse varies on average between 20 and 30 days (Fig.3). The metacyclic forms are transmitted to the HD during a subsequent bite by the parasitized fly.

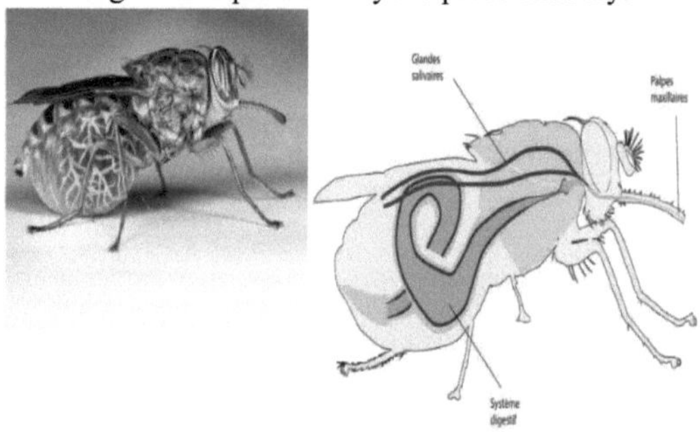

Figure 4: Tsetse fly after a blood meal and its schematic representation

- Tsetse flies are strictly African and transmit sleeping sickness, also known as **African trypanosomiasis**. So there are:

- **West African trypanosomiasis**, which occurs in West Africa and is caused by Trypanosoma brucei gambiense, transmitted by Glossina palpalis
- **East African trypanosomiasis**, caused by Trypanosoma brucei rhodesiensee and transmitted by Glossina morsitans.

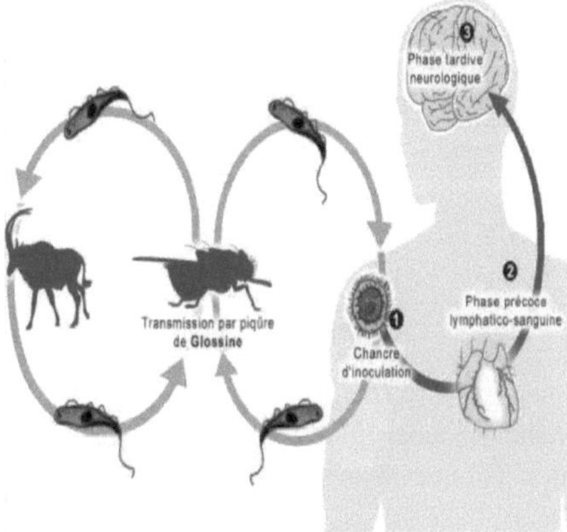

Figure 5: Transmission cycle African trypanosomes and human/animal African trypanosomiasis

Transmission of trypanosomes by an infected tsetse fly on a healthy human or animal host. At the site of the bite, inoculation chancre may be observed (Fig.5). The early phase of the disease is the haemolymphatic phase and the second phase is the meningoencephalic phase.

II.2. Species: Leishmania spp, agent of leishmaniasis

- Kingdom: Protista
- Phylum of Sarcomastigophora (Rhizoflagellates)
- Subphylum Mastigophora (Flagellates)
- Class of Zoomastigophora (or Zooflagellates)
- Order Kinetoplastida
- Suborder Trypanosomatina
- Trypanosomatidae family
- Genre: Leishmania

Leishmaniasis is an anthropozoonosis or anthroponosis caused flagellate protozoa belonging to the genus Leishmania. This parasite is **di-morphic**: it is carried by a female of a small dipteran insect, the sandfly, in the form of a mobile flagellated promastigote, and once inoculated into a mammal, this parasite differentiates within the phagocytic cells into a rounded intracellular form known as an amastigote. The vectors of leishmaniasis are female haematophagous sandflies of the order Diptera, genus Phlebotomus. There are nearly 1,000 known species of sandfly, of which fewer than 50 are vectors of the Leishmania parasite.

• **Morphology :**

- Ovoid, 2 to 3 µm in diameter; 1 nucleus + 1 kinetoplast. Multiplication by scissiparity, immobile, necessarily endocellular (Fig.6a).

- **Promastigote form** in the insect intermediate host (Phlebotomus): elongated, flagellate, 8-24 x 4-5 µm long and highly motile, moving flagellum forward (Fig.6b); multiplication by scissiparity. Found in the digestive tract of intermediate hosts and in in vitro culture.

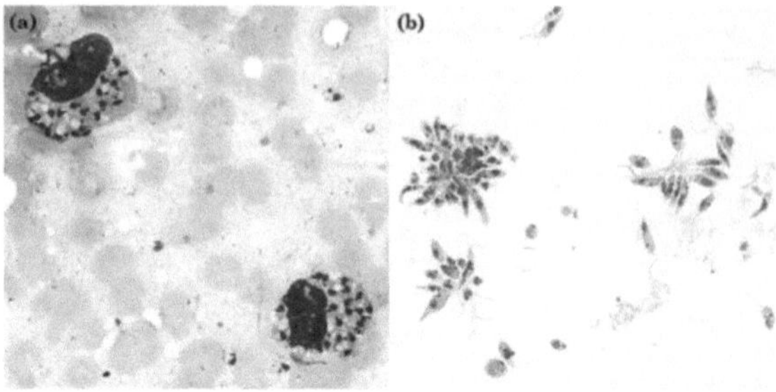

Figure 6: (a) Leishmania sp. amastigotes in two macrophages with Giemsa-stained preparation. (b) Leishmania sp. promastigotes from culture.

- **Evolutionary cycle:** This parasite occurs in two distinct forms: the amastigote form HD and the promastigote form HI.

- **Definitive host (HD): humans and animals** (dogs, wild rodents).

In the vertebrate host, the parasites are located in the skin and blood and are intracellular and localised in the cells of the mononuclear phagocytic **system** or **reticuloendothelial system.** They multiply by scissiparity, allowing the parasite to spread within host. This form of the parasite, known as **amastigote,** is ingested by sandflies during a blood meal (Fig.7)

-Intermediate host (HI): a biting insect of the sandfly group **(genus: Phlebotomus), the female of** which **is a haematophagous**. Phlebotomines are nocturnal and crepuscular, and fly silently. They are 2 to 3 mm long and can pass through the mesh a mosquito net.

In the vector, amastigote leishmania are rapidly transformed into procyclic promastigotes over the following 24 to 48 hours (Fig7). These mobile extracellular forms an anteriorly positioned flagellum multiply for 3 days and eventually cross the membrane into the intestinal lumen where they migrate to the foregut where the parasites attach to the microvilli of the epithelium. At this stage, the procyclic leishmanial promasigotes can pass into the metacyclic phase, which is the infesting form for the hosts. At the end of the cycle in the vector, the metacyclic forms collect in the fore midgut of the sandfly (proventriculus or pharynx, depending on the subgenus), ready to be injected into a new vertebrate host during the next blood meal. In total, the time required for the parasites to complete the cycle in the vector varies between 1 and 2 weeks, depending on the species.

Metacyclic promastigote parasites are transmitted to the host during the sandfly's second blood meal. Once the parasites have been injected into the dermis of vertebrate hosts, their establishment is facilitated by a number of adaptive mechanisms, such as the injection of saliva by the vector, which plays a role in the success of the infection (Kamhawi et al. 2000), or the implementation of apoptosis in leishmania, which favours the colonisation of macrophages (Van Zandbergen et al. 2004).

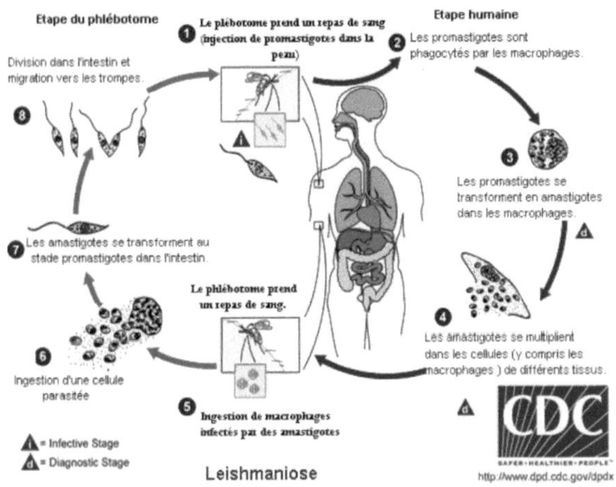

Figure7: Complete parasitic cycle of Leishmania in hosts and vector. Adapted from Harhay et al (2011).

The parasites are then phagocytosed by macrophages and lose their mobility again, returning to the amastigote form to multiply in a parasitophorous vacuole (Fig.7). This vacuole is formed by the parasites as a new cellular compartment allowing them to divide without being digested by macrophages. Transformation into endocellular amastigote forms takes place in a few minutes.

- **Clinical forms**

The clinical manifestations of leishmaniasis depend on the infecting species of Leishmania, the biology of the vector and the immune response of the host (Tab.1). The result of these interactions is a spectrum of disease that can range from localised skin lesions to diffuse damage to the reticuloendothelial system. Traditionally, human leishmaniasis has been classified into three main clinical syndromes: visceral leishmaniasis (VL) or Kala- azar, cutaneous leishmaniasis (CL) and mucocutaneous leishmaniasis (MCL). However, there other variants of the disease, given the great diversity of the Leishmania species and the immune response, which differs from one host to another.

Table 1: Clinical forms of leishmaniasis in Algeria according Leishmania species and vector

Agent pathogen	In the world	Agent pathogen	Vector	Algeria
L. donovani	LV	L. infantum	P.perniciosus p.longicuspis	LV LC dry form
L. tropica	LC	L. major	p.papatasi	LC Zoonotic or Biskra nail : wet form
L. braziliensis	LCM	L. killicki	P.sergenti	LC

VL is the most serious form of leishmaniasis, since if left untreated, it can lead to the death of the patient.

The disease can be life-threatening. Clinical manifestations are dominated by irregular and prolonged outbreaks of fever, splenomegaly, anaemia and loss of blood.of weight. Their intensity reflects the multiplication and establishment of parasites in certain organs such as the bone marrow and spleen. VL used to be considered a paediatric disease, and it was not until the 1980s that adult cases of VL were described, particularly in immunocompromised subjects. Two strains of Leishmania are known to cause VL: L. donovani, the anthroponotic species, which is highly endemic in East Africa and the Indian subcontinent, and L. infantum, the zoonotic species (reservoir: dog), which is found around the Mediterranean, in the Middle East and in South America.

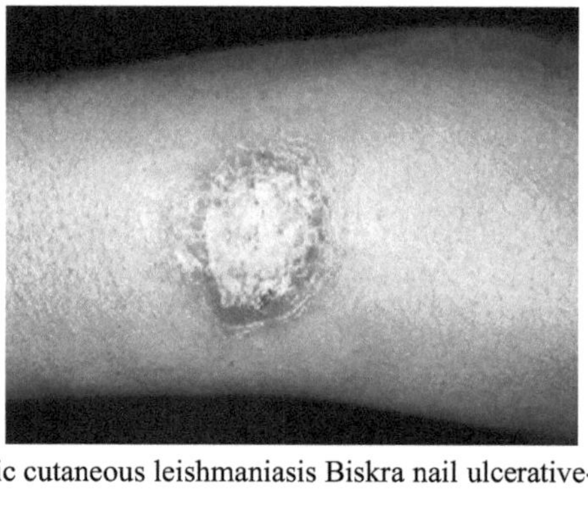

Figure 8: Zoonotic cutaneous leishmaniasis Biskra nail ulcerative-crust appearance

Cutaneous leishmaniasis (CL) is the most widespread form of leishmaniasis in the world. Inoculation of the skin with Leishmania promastigote causes a skin lesion that heals spontaneously but leaves unsightly scars (fig.8). Mucocutaneous leishmaniasis (MCL) involves total or partial destruction of the mucous membranes of the nose, mouth and throat. This condition is known as "It begins the appearance skin lesions perfectly similar to those of LC, which eventually heal in 6 months to 1 year. These lesions are followed (or concomitant) by another infection that attacks the nasal and oral cavities preferentially. Ulcerative lesions may extend to the oropharynx and trachea and can be very disfiguring. This pathology is essentially caused by the L. braziliensis strain.

- Kingdom: Protista
- Phylum of Sarcomastigophora (Rhizoflagellates)
- Subphylum Mastigophora (Flagellates)
- Class of Zoomastigophora (or Zooflagellates)
- Order of Diplomonadida
- Suborder: Diplomonadina
- Genre: Giardia
- Species: Giardia intestinalis

Giardia duodenalis, Giardia lamblia or intestinalis is a flagellate protozoan that occurs in two forms: vegetative or trophozoite and cystic.

1. Vegetative form: The body is kite-shaped, measuring 10 to 20 microns by 6 to 10 microns, with bilateral symmetry in relation to a median axis represented by the axostyle (Fig.9). This shape is concave ventrally and convex dorsally. It has 2 nuclei, 2 comma-shaped parabasal bodies and 4 pairs of flagella: 3 anterior pairs and one pair at posterior end. It lives in the duodenum and the beginning of the jejunum. It makes rapid movements reminiscent of falling leaves, turning on itself and attaching itself by suckers to the base of the villi of the small intestine. This form is rarely seen under coproscopy, except when fresh stools are examined directly.

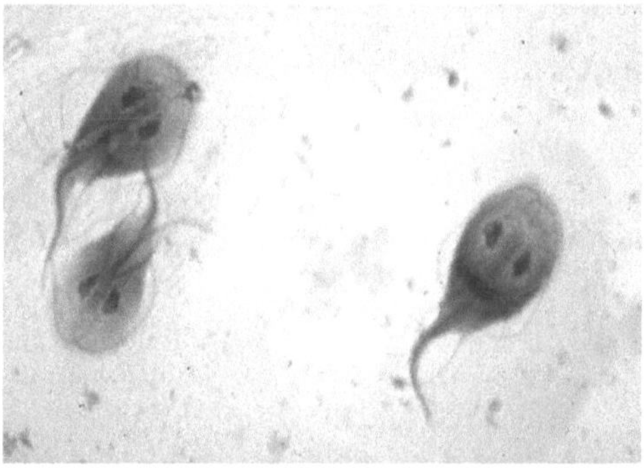

Figure 9: Giardia intestinalis in vegetative form

2. The cystic form: Ovoid, 8 to 12 microns long and 7 to 10 microns in diameter, with a thin, highly refractive shell and two comma-shaped parabasal bodies. Flagellar residues are grouped in a bundle along the axis of the cyst. Pre-cystic forms have 2 nuclei, mature cystic forms have 4 nuclei. This form develops in the colon. Cysts are very resistant, especially in water, and can survive at low temperatures (2 months at+ 8°C). Bleaching water at the concentration usually used to sterilise drinking water is insufficient to kill them, but boiling and freezing (at -20°C) destroys them.

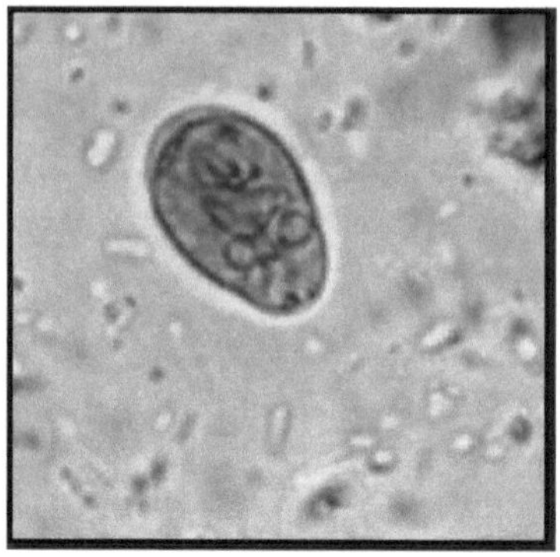

Figure 10: Giardia intestinalis in cystic form

Evolutionary cycle :

The life cycle of Giardia is monoxenous and comprises only two stages, the encystation stage and the trophozoite production stage (Fig.11). Most of the cycle takes place inside the host, whether human or animal, which triggers excystation when ingests the cysts. Combined exposure acid in the stomach and bile salts in proximal small intestine stimulates the release of trophozoites from the cysts, which then attach to and colonise the mucosal surface of the small intestine, where they rapidly multiply by binary asexual fission. Trophozoites in small intestine progressively form cysts which are eliminated in the faeces, and this is where the exogenous phase of the cycle begins. Outside, these cysts remain quiescent if they find the humidity and temperature favourable to their maintenance in the environment. They are resistant and persist for a long time in water, mud, soil and food they are ingested by another host.

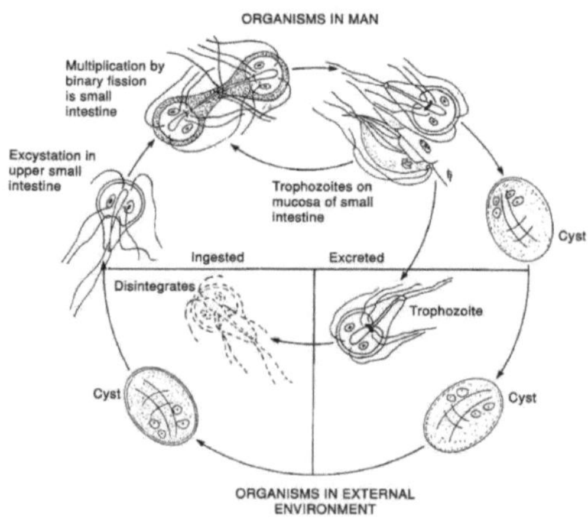

Figure11: Evolutionary cycle of Giardia intestinalis (Smith and Paget, 2007)

Transmission modes

Transmission occurs through **the ingestion of cysts:**
- Giardiasis can be transmitted by drinking contaminated water and, more importantly rarely, by eating contaminated food.
- transmission also occurs through direct or indirect contact following the cyst contamination of the environment and hands.

Epidemiology

G. intestinalis is a ubiquitous parasite found in temperate and tropical regions, with prevalence varying according to age and socio-economic level. Although it is common, most sufferers remain asymptomatic. In any case, the symptoms of giardiasis vary, depending on the strain, the environment (frequency of transmission) and the immune response of the host organism.

Clinic

Infection is asymptomatic in the majority of children and adults. Symptomatic infection occurs more frequently in children. It is associated with a wide variety of symptoms: acute, chronic or intermittent diarrhoea, fatigue, bloating, flatulence, abdominal cramps, greasy, pale and foul-smelling stools, loss of appetite, weight loss, vomiting and fever. The infection is benign in most .

Complications

Very rarely, chronic infection may lead to malabsorption syndrome with vitamin deficiencies and lactase deficiency (lactose intolerance), which may persist for several weeks after treatment. Children under 5 years of age, pregnant women and immunosuppressed patients may develop a more serious infection (e.gdehydration requiring hospitalisation).

Biological diagnosis

1. Microscopic investigation: - parasitological examination of the stools, repeated 3 times, several days apart, because there are no periods of transmission (cysts on direct examination or after concentration: MIF, Ritchie, Bailenger),

- examination of duodenal fluid taken by Enterotest or aspirated during digestive endoscopy (trophozoites),

- endoscopic duodenal biopsies (trophozoites in the intestinal lumen).

2. Rapid diagnostic strip **tests** for the detection of Giardia intestinalis antigens in stools. This is an acute screening test.

3. Use of antibodies monoclonal antibodies for the detection of parasites by immunofluorescence or detection of copro-antigens by immunofixation or ELISA

4. Serological diagnosis: determination of serum IgG by ELISA, which is useful studying epidemiology.

5. Advantages real-time PCR: it can detect the parasite and determine the specific genotype

Prophylaxis

Giardiosis is a disease caused by faecal peril: prophylaxis is the same as for faeco-oral infections, particular hand hygiene and health education.

- Kingdom: Protista
- Phylum of Sarcomastigophora (Rhizoflagellates)
- Subphylum Mastigophora (Flagellates)
- Class of Zoomastigophora (or Zooflagellates)
- Order Trichomonadida
- Genus Trichomonas
- Species : T. vaginalis

Urogenital trichomoniasis is a common, cosmopolitan, benign sexually transmitted infection (STI) caused by Trichomonas vaginalis, a flagellate protozoan, a strict parasite of the urogenital tract. It is the 1st cause of sexually transmitted infection in the world. Trichomonas vaginalis is frequently found in women between the ages of 16 and 35, and affects around 20-25% of them during the fertile period. Vaginitis caused by Trichomonas vaginalis accounts for 50% of vaginitis with leucorrhoea.

2. Morphology

Trichomonas vaginalis is a strictly human parasite that exists only in vegetative form and dies rapidly in the external environment. There is no cystic form.

The vegetative or trophozoite form

This shape is mobile, almond-shaped, rounded or oval, colourless, refractive when fresh under the microscope, and measures 10 to 15µm long by 7-10µm wide. It has an axostyle that crosses the cell and protrudes behind it, an oval nucleus in the anterior part of the body, a ribbon-shaped parabasal body curved over the nucleus and a kinetoplast, from which four free anterior flagella and a recurrent flagellum emerge and attach to the body of the parasite forming a a wavy membrane about half the length of the cell. Near the ovular nucleus is a band-like element, the parabasal body (Fig.12).

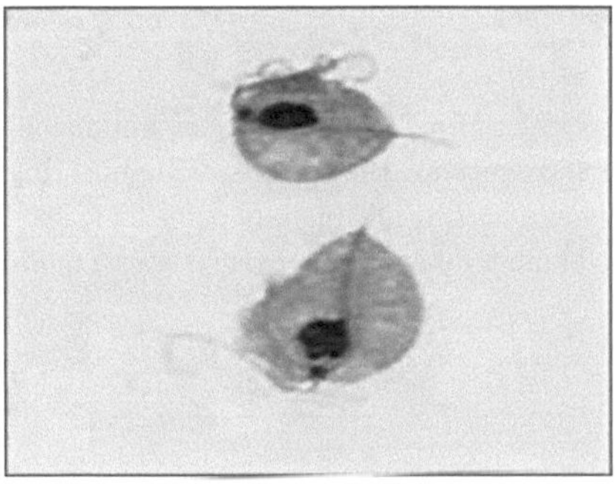

Figure 12: vegetative form of Trichomonas vaginalis

3. Habitat

 In women, after puberty and when vaginal acidity is reduced, it is frequently found as a parasite of the vaginal cavity and the urethra, but also of the bladder and the Skene and Bartholin glands, which are thus untreatable landmarks and a possible source of recurrence.

 In men, it is found in the urethra, balanopreputial fold, prostate and seminal vesicles.

4. Biology

Highly sensitive to desiccation, it can survive for 1 to 2 hours on a damp surface and up to 24 hours in urine or semen at 20°C. Optimum conditions for growth are a temperature of 35-37°C, a pH of 5.5-6 and anaerobiosis. Actively moving, it feeds by osmosis and multiplies by longitudinal division on the surface of the urogenital mucosa men and women.

5. Pathogenesis

Despite the existence of many healthy carriers (50% of cases are asymptomatic), especially in men, Trichomonas vaginalis has a clear pathogenic role in both sexes: it is responsible for 30% of non-gonococcal urethritis. It is a sexually transmitted disease, the risk of transmission of which depends on age and sexual activity.

B. Definitive host/ Parasite reservoir

Humans are the only host and reservoir of the parasite.

C. Mode of contamination

Contamination is only by direct sexual contact. Some authors admit extra-sexual contamination via shared toiletries (mother-daughter). However, this type of contamination is exceptional, as Trichomonas vaginalis does not live longer than half an hour to an on a piece of toilet paper.

Evolutionary cycle

The fragility of the vegetative form of the parasite and inability to encyst mean that transmission is only direct and venereal. The cycle is single-host (Fig.13), passing from one the infected subject directly to a healthy and directly pathogenic subject. Multiplication occurs by scissiparity (longitudinal binary division); there is no sexual multiplication.

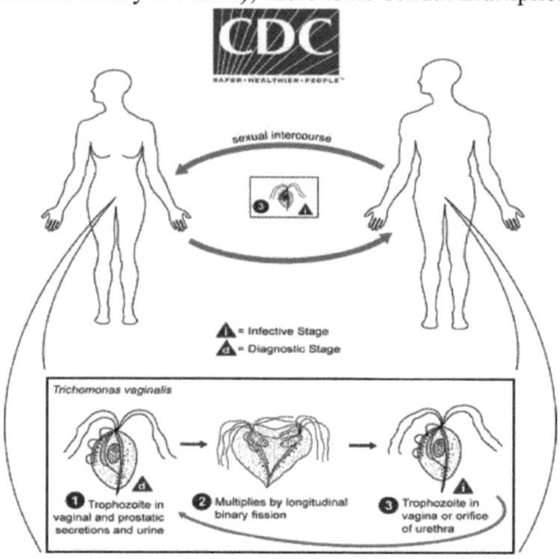

Figure 13: Evolutionary cycle of Trichomonas vaginalis.

Contributing factors

- Hormonal disorders caused by oral contraception, the menopause and pregnancy lead to a hormonal imbalance.
- The alkalinity of the pH of vaginal secretions (the normal pH of the vagina is between 3.8 and 4.5).
- Associated flora: staphylococcus, streptococcus, lactobacillus.
- Humidity.

Clinical signs

- Men are generally healthy carriers, asymptomatic or only slightly symptomatic.
The rate of transmission sexual intercourse is 80%.
- In women, continuous, foul-smelling, foamy, aerated, greenish-yellow (sometimes whitish) leucorrhoea associated with vulvar pruritus with a burning sensation, dyspareunia and sometimes cystitis (dysuria, pollakiuria, urinary burning) are the main signs that often lead women to seek medical attention. On examination, the vulva is bright red with a painful exudate when the speculum is inserted. The vaginal mucosa is scarlet red with darker haemorrhagic staining.

- In men, there may rarely be a genital discharge, morning drop, cystitis or haemospermia. The absence of signs in men encourages the spread of Trichomonas vaginalis.

Biological diagnosis

1. Direct debit

□ **In women**: Vaginal secretions are collected after inserting a speculum into the posterior cul de sac, before any intimate hygiene or treatment. The patient should avoid sexual intercourse for 24 to 48 hours prior to sampling, which should be carried out at the level of the vaginal cul-de-sac and Bartholin's gland, using a Pasteur pipette or a sterile swab soaked in saline .

□ **In men**: The sample is taken before any morning urination. The first morning serum is collected from the meatus and the urine from the first jet. Massaging the prostate increases the sensitivity of the sample. If the sample is taken outside the laboratory, a swab with a transport medium (Stuart medium) must be used, which allows the parasites to be preserved for 24 hours at room temperature.

Trichomonas vaginalis can also be tested in the centrifugation pellet of freshly passed urine.

2. Direct examination

The urine exudate or centrifugation pellet is examined between slide and slide to look for motile forms of T. vaginalis under a light microscope with an X10 or X40 objective. Direct examination should be carried out as quickly as possible avoid cooling the vegetative forms of T. vaginalis, in drop of physiological water at 37°C. This examination can be carried out after staining a dried smear then stained with May Grunwald Giemsa (MGG).

Immunological diagnosis

T. vaginalis antigens found in all isolates can be detected monoclonal antibodies labelled peroxidase or fluorochrome. Detection can be carried out a variety of methods, including haemagglutination, immunofluorescence and enzyme linked immunosorbent assay (Elisa).

D. Molecular diagnosis

PCR methods can detect target sequences with commercially available oligonucleotide probes with a sensitivity of 95% and a specificity of 98%.

Prophylaxis

□ Observe hygiene , in particular the strictly personal use of intimate toiletries.

□ Treatment of partner(s) and standard venereology precautions;

- Kingdom: Protista
- Phylum of Sarcomastigophora (Rhizoflagellates)
- **Subphylum Sarcodina** (Rhizopoda)
- Superclass Rhizopoda
- Order Amoebida
- Family Entamoebidae

There are 8 species of amoeba that parasitise the human colon and belong to 4 different genera, five which belong to the Entamoeba genus: Entamoeba coli Entamoeba hartmani Entamoeba polecki Entamoeba dispar Entamoeba histolytica Other species include:
Endolimax nanus Pseudolimax butschlii. Dientamoeba fragilis
One species is found in mouth: Entamoeba gingivalis Of these species, only **E. histolytica** is pathogenic for humans **Morphology of parasitic amoebae in humans**
Amoebae are microscopic, unicellular organisms of irregular shape, moving with the aid of pseudopods (Rhizopodes). They consist a single cell, with a cytoplasm that is sometimes divided into ectoplasm and endoplasm. Identification of the different species is based on the microscopic appearance of the nucleus, which is rounded and contains a chromatin corpuscle of varying size called a karyosome.

-In the Entamoeba genus: the karyosome is small, punctiform and the chromatin is peripheral, attached to the nuclear membrane called the chromatin croutelle.

-In other genera, the karyosome is the only component of the nucleus, and is a large central or eccentric karyosome. Amoebae are found in two forms:

- Vegetative form.
- Cystic form.

■ **Entamoeba histolytica (dysenteric amoeba), agent of amoebiasis**

Amoebiasis is a parasitosis caused by Entamoeba histolytica, primarily localised in the intestine and which may migrate secondarily to other organs (liver, lung, brain, etc.). The WHO defines amoebiasis as a pathological condition in which the body harbours Entamoeba histolytica with or without clinical manifestations.

Entamoeba histolytica can present itself in 3 different morphological forms: 02 vegetative forms or trophozoites and a cystic form.

■ **Trophozoite form:** vegetative and multiplication cell form (Fig14a).

a/ Trophozoite form "minuta aspect": 10 to 15 µm, peripheral nucleus of 3 to 4 µm with a small central karyosome and peripheral chromatin in edged or dotted form. When fresh, movements are brisk and the pseudopods are long; optimum temperature is below 37°C. This is the saprophytic, non-haematophagous and non-pathogenic form, as it feeds on bacteria and food debris. It lives in the lumen of the colon or on the surface of the colonic mucosa.
b/ Trophozoite form "histolytica aspect": same general appearance as the minuta form, but larger in size than the minuta form (20-30 µm to 40 µm) with the presence of additional characteristics; it contains in its cytoplasm red blood cells in the process of digestion, is the haematophagous and pathogenic form that feeds on red blood cells. It lives on the colonic mucosa and can invade other organs. It tolerates 37°C well. These two trophozoite forms multiply by binary division.

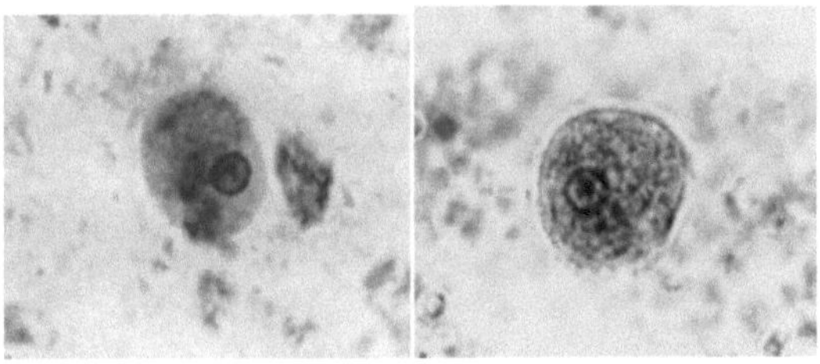

Figure 14a: Entamoeba histolytica "vegetative form

Figure 14b: Entamoeba histolytica "vegetative form

The cystic form: this is the form of passive dissemination of the disease and resistance in the external environment (Fig14b). It is spherical, 10 µm to 15 µm in diameter and surrounded by a rigid shell. Immature cysts contain a vacuole, 1 or 2 nuclei and are larger. When mature, they contain 4 nuclei and sometimes round-tipped rods called siderophilic or crystalloid bodies. They are eliminated in the faeces of patients and healthy carriers and are highly resistant in the external environment. In the outside environment, these cysts have a variable survival time depending on humidity and temperature: around ten days in damp ol at 18°C, 3 months in water at 4°C. Heat (50°C) and desiccation kill them, but they are resistant to low concentrations of bleach.

Evolutionary cycle

a) Non-pathogenic cycle.

Ingestion of mature cysts with 4 nuclei is followed by de-cysting in the gastrointestinal tract (Fig.15). The nuclei divide once, giving eight minuta-type amoebules which multiply by binary division on contact with the colonic mucosa and are eliminated into external environment in the form of faecal cysts.

b) Accidental pathogenic cycle (Passage to pathogenicity and amoebiasis disease): Given the right conditions (host factors+ parasite factors= pathogenic potential of strains), the trophozoite minuta forms transform into histolytica. These grow, erode the intestinal wall and become haematophagous (Fig.15). They have significant lytic power over tissues and progress from person to person, creating extensive ulcers in the intestinal wall (shirttail abscesses). No cysts form during this phase, so there is no direct epidemiological role.

- The parasite returns to the minuta form after 3 weeks ==> **end of the amoebic crisis.**
- Possibility of blood and/or lymphatic metastases from a colonic abscess, leading to extra-colonic localisations and formation of amoebic abscesses in other organs: liver (most frequent form), lungs, more rarely brain.

In extra-colonic cases, there is no return to the minuta form.

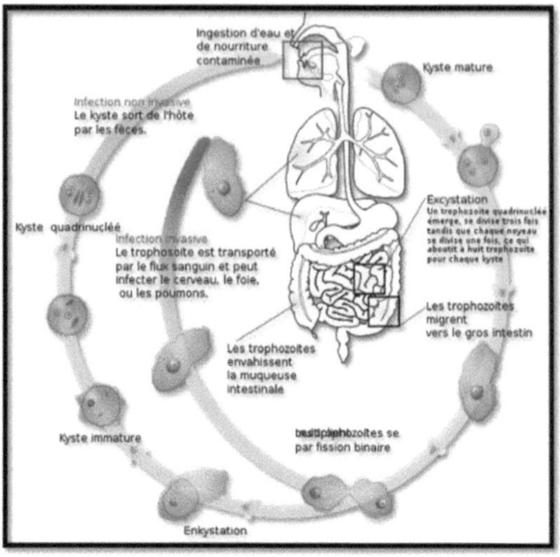

Figure 15: Entamoeba histolytica cycle

- **Epidemiology**

Amoebiasis **is a cosmopolitan** disease. It is highly prevalent in tropical environments, due to poor faecal hygiene. In temperate countries, it is found among immigrants and tourists who have spent time in highly endemic countries, as well as residents of institutions for the mentally ill and homosexuals, both HIV-negative and HIV-positive.

- **Mode of contamination**

The usual mode of contamination is indirect contamination through food (raw vegetables) or drinks that have undergone faecal contamination. The cysts, which are resistant in the external environment, are then the infesting forms. Another possible mode of contamination is direct contact, in the case of human concentrations: prisons, asylums, boarding schools, camps, etc.

- **Pathophysiology**

a) Pathophysiology acute intestinal amebiasis

Humans become infected by ingesting cysts, which transform into trophozoites in the digestive tract. These Entamoeba histolytica trophozoites adhere to the colonic wall via lectins. Affected human cells are killed and destroyed within minutes by the formation of pores in their membrane. The production of proteolytic enzymes (cysteine proteinases) by the amoebae favours their diffusion into the colonic mucosa and sub-mucosa, leading to oedematous thickening, the formation of multiple ulcerations, areas of necrosis and sometimes intestinal perforation.

Endoscopic examination of the colon reveals suggestive "nail-bite ulcerations" and "shirt-button abscesses". Further infection and possible spread depend in part on the host's local immune response.

b) Pathophysiology of hepatic and tissue amebiasis

invasion of the colonic wall, amoebae can cause disruption of the micro-vascularisation and spread by haematogenous route into the portal system. The amoebae then adhere to the walls of the hepatic capillaries and destroy the hepatic parenchyma centrifugally, resulting in an amoebic abscess of the liver. Hepatic localisation is always secondary to colonic contamination, but it may appear at a distance from the episode. dysenteric.

■ **Clinic**

a) Amoebiasis-infestation

Presence of amoeba "minuta" in the colon and cysts in the stools, **with no pathological manifestations (healthy carriage).**

b) Amoebiasis disease

b-1) Acute intestinal amoebiasis

It manifests itself mainly in the form of amoebic dysentery (frequent evacuation of non-fecal stools):

- Severe abdominal pain.
- Bloody mucous sputum : Five to six "bowel movements" / day
- Tenesmus is a painful contracture of the anal sphincter accompanied by an urge to defecate.
- The general condition is well preserved and apyrexia is the rule.

- If the dysenteric syndrome persists: asthenia, weight loss, greyish complexion ("potato complexion") and may be complicated by intestinal haemorrhage with risk of peritonitis due to the necrotising action of the amoeba.
- The amoeboma is an inflammatory parasitic pseudotumour that occurs preferably in the cecum, and manifests clinically as an intestinal subocclusion syndrome.

b-2) Tissue amoebiasis (Extra intestinal)

- **Hepatic amebiasis:** amebic abscess of the liver, the most common form. It results from the hepatic localisation of the "histolytica" form, which migrates from abscesses in the colonic mucosa to the liver. At the focus, there is parasitic thrombosis, followed by hepato-necrosis; the foci enlarge, coalesce and form an amebic abscess clinically characterised by (Fonton's triad):

- pain in right hypochondrium
- Fever
- Hepatomegaly, which may accompanied a sub-icteric or state.
- The abscess (on puncture) contains sterile chocolate-coloured pus.
- Parasitological examination of stools is often negative at this stage, in the absence of dysenteric syndrome.
- **Pulmonary amebiasis**

The lung is the organ most frequently invaded after the liver. The following may be observed:

- Pleural effusion or purulent amebic pleurisy.
- A lung abscess, usually in the right lower lobe.
- Bronchoalveolar fistula.

Other locations are possible: brain, spleen, etc.

Biological diagnosis

1- Presumptive diagnosis

Epidemiological context: stay in endemic zone: hot, humid countries (favouring factors: humidity, T°, socio-ecological conditions)

2- Clinical diagnosis: Imaging

3- Non-specific biological signs: hyperleukocytosis, increased ESR, increased transaminase, sometimes jaundice with obstruction.

blood hypereosinophilia

-Rectoscopy: painful but essential, inflammatory mucosa is seen.

4- Direct parasitological diagnosis :

- **Acute intestinal amebiasis**

Search for vegetative forms or cysts: samples taken 3 or 4 days apart. Afecal, mucusy, bloody stools. Ideally, the patient should be on a low-residue diet (fish, white meat, eggs, pasta and rice) 3 days before stool collection.)

- **Hepatic and pulmonary amebiasis :**
- Biopsy: the chocolate pus is examined.
- Acid tests (ELISA, haemagglutination, IFI) indirect immunological diagnosis is only used in deep forms.
- Search for vegetative forms: stool suspensions: o without staining: histolytica form visible o with MIF staining (merthiolate--iode--formol): differentiation of amoeba types, visualisation of nuclei Stool concentration (BAILANGER technique; modified RITCHIE technique)
- Differential diagnosis: with Entamoeba coli or E. dispar

- Culture: NNN medium (Novy--Mc Neal--Nicole)
- Coproculture on two-phase agar medium+ starch: Dobell and Laidlaw medium

Immunological diagnosis: Search Ac: For tissue forms because of the presence Ac unlike colonic forms where the immune response is minimal. Use of 2 techniques followed by a confirmation test:

--IFI --Immunoelectrophoresis, electrosynthesis --indirect hemagglutination --agglutination on sensitized latex particles: BICHROLATEX --ELISA Search for copro--Ag: possible by ELISA and IFI colonic forms

■ **Prophylaxis: Combating faecal peril**

▶ **Collective**

Waste water treatment,

Prohibit the spreading of faecal matter Protect market garden crops Disinfection of patients' faeces

Detect and treat healthy carriers, treat chronic amoebic patients

▶ **Individual**

Washing fruit and vegetables.

Avoid foods that irritate the intestine.

Washing your hands. Boil thc water

■ Phylum of Apicomplexa (sporozoa)

■ Sporozoea class

■ Coccidia subclass

Apicomplexa (sporozoa) are protozoa characterised by the presence of an apical complex comprising a cytoskeleton and 3 types organelles: miconemes, rhopties and dense granules.

- **Micronemes** are small rod-shaped organelles located in the anterior half of tachyzoites and bounded by a membrane. There are around fifty of them and they play the same role as the rhoptries.
- **The rhoptries**, of which there are about ten or twenty, are club-shaped, 1 to 4 µm long, and are located in the anterior third of the parasite. Their anterior ends grouped together in two ductules to form an apical vesicle. They have a secretory function, a proteolytic enzyme which plays a role in the cell penetration mechanism [53].
- **Dense granules** are cytoplasmic organelles 200 µm in diameter, located on either side of the nucleus.

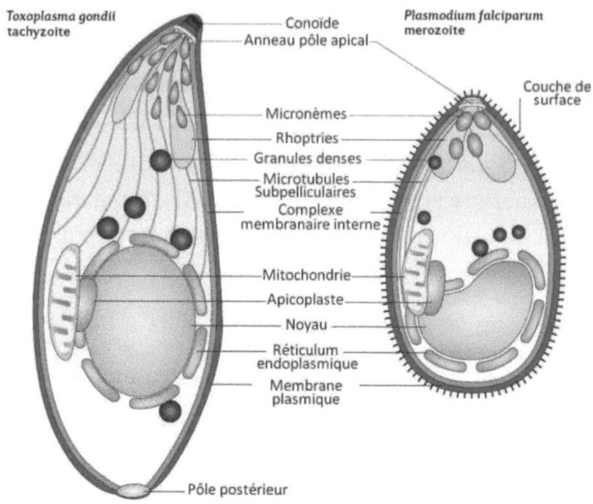

Figure 16: Apicomplexa (Sporozorya) apical complex structure

- Phylum of Apicomplexa (sporozoa)
- Sporozoea class
- Coccidia subclass
- Eimeriina sub-order
- Family Sarcocystidae
- Genus Toxoplasma
- Species : T. gondii

Toxoplasma gondii, the agent of toxoplasmosis

Toxoplasmosis is a very common cosmopolitan zoonosis caused by obligate intracellularly multiplying protozoan Toxoplasma gondii. It is generally asymptomatic, although in immunosuppressed individuals and foetuses infected during the first trimester of pregnancy, the disease can be severe. Toxoplasma gondii is a facultative heteroxenous coccidia (monoxenous or heteroxenous cycle), with obligatory intracellular development. It undergoes three infectious stages during its life cycle:

• **Tachyzoites**: formerly called "trophozoites" and derived from the Greek word "tracus", which indicates the rapidity of cell division in an intermediate host cell. It is the rapidly multiplying asexual vegetative form and is the only form that can cross the placental barrier. Morphologically, the tachyzoite has the shape a crescent 6 to 8 microns long and 3 to 4 microns wide (Fig.17). They multiply in the blood, biological fluids and tissues, are fragile and destroyed gastric acidity, so are not infectious by mouth, but are infectious by blood to the foetus in congenital toxoplasmosis. They are present only during the acute phase infection.

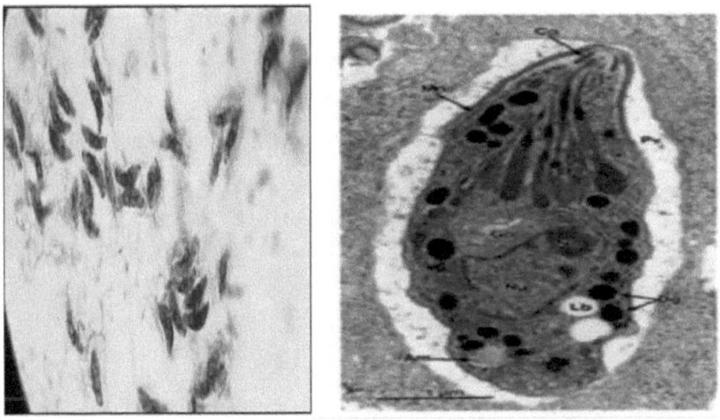

Figure 17: "Left" Tachyzoites seen under the electron microscope. "Right: Ultrastructure of a Toxoplasma gondii tachyzoite; Co: conoid; Mn: microneme; PV: parasitophorous vacuole; Rh: rhoptrie; Go: Golgi; No: nucleolus; Nu: nucleus; Lb: lipid body; Dg: dense granules; Am: amylopectin.

• **Bradyzoites:** This is the form of tissue resistance resulting from the transformation of tachyzoites by several asexual multiplications. The term bradyzoite comes from the Greek word

The word "bradus" signifies the slowness of their division within cysts in the host cell. Morphologically, they are spherical or ovoid in shape, measuring between 50 and 50 mm in diameter. 200 µm, contain thousands of bradyzoites that are found mainly in neurons, astrocytes, muscle cells and retinal cells (Fig 18). These cysts can survive for several days at room temperature and several months at 4°C. They are destroyed by heat (a quarter of hour at 56°C) or freezing (24 hours at -20°C). This is one of the methods of contamination humans by oral route through ingestion of parasitized meat.

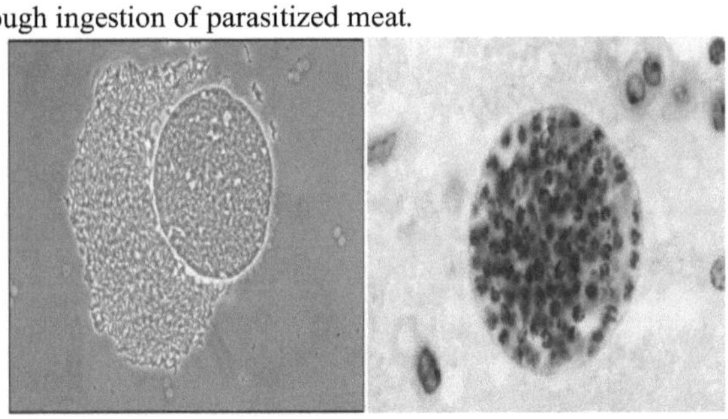

Figure18: "right" T. gondii tissue cyst in the brain of a mouse. "Left: Rupture of the cyst wall and release of hundreds of bradyzoites under the action of digestive juices.

• **Sporozoites:** This is the form of resistance in the external environment. They are found sporulated oocytes (Fig. 19). Oocysts ovoid structures 9 to 11 µm long and 11 to 14 µm wide, resulting from sexual reproduction in the intestinal cells of the definitive host: the cat. They consist of 2 sporocysts containing 4 sporozoites each (a sporozoite resembles a tachyzoite).

This is the parasite's form of resistance in the external environment. Oocytes can survive more than a year in damp soil and are the main form of contamination of herbivores and humans through the ingestion food contaminated by cat faeces.

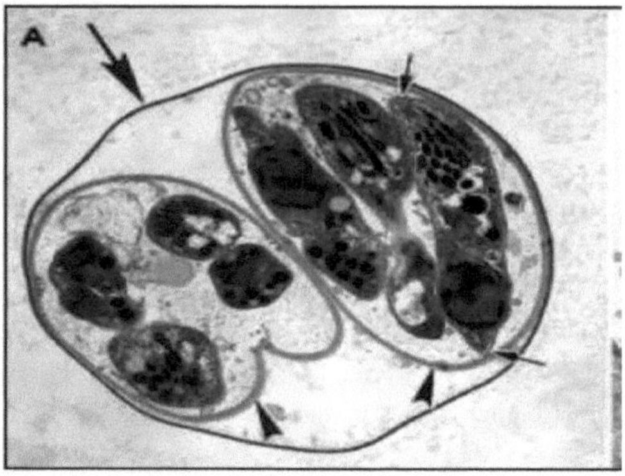

Figure 19: Toxoplasma sporulated oocyst observed by electron microscopy (Dubey, 1998). (A) Sporulated oocyst protected by its wall (large arrow) containing two sporocysts (arrowheads) each containing four sporozoites (small arrows).

Epidemiology of Toxoplasmosis: Toxoplasmosis is a cosmopolitan zoonosis with a worldwide distribution. Its prevalence is linked to several factors [102,104]: climatic differences, level hygiene, dietary habits and diet, quality of drinking water, age and the presence of felids.

Main sources of human contamination :

- ingestion of oocysts through telluric walls (water, food, hands) 2% of cats are oocyst sowers, for 1 to 3 weeks (especially young cats) sporulation conditions: 1 to 5 days at 20°C in the presence oxygen; oocysts survive for several months on soil, and are also resistant to HCl. They are killed by heat, desiccation and freezing.

- ingestion of live cysts in raw or undercooked meat (Fig.20); cyst survival: several days at room temperature, several months at+ 4°C destroyed by : 15 mm at 56°C, 24 hours at - 20°C

- Transplacental passage
- Transmission by organ transplant, transfusion

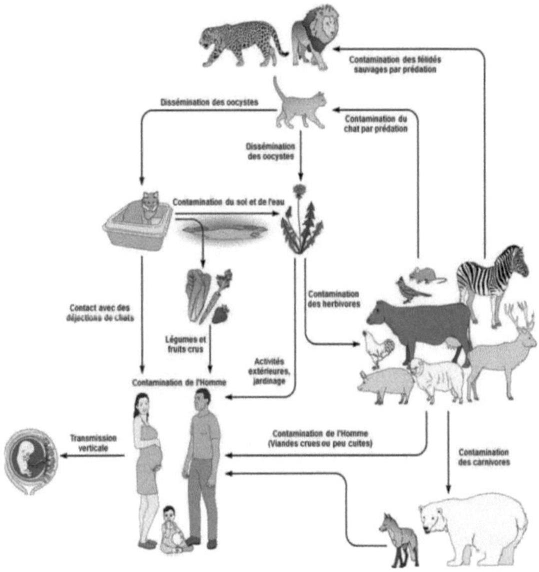

Figure 20: Sources of Toxoplasma gondii contamination

Evolutionary cycle of Toxoplasma gondii

There are two cycles (Fig 21): a sexual cycle that occurs only in felids, and an asexual cycle intermediate hosts: humans or herbivorous animals such cattle (sheep and cattle) or omnivores (pigs, rats, etc.).

- The cycle is tenfold when the definitive host and intermediate hosts are involved.
- The cycle is monoxenous if the toxoplasma is transmitted from intermediate host to intermediate host without infesting the definitive host. The cycle then takes place without sexual reproduction. [80]

- **Definitive host (HD):** felids

- Location: digestive tract=> asexual then sexual reproduction=> oocysts in external environment: maturation of oocysts containing 2 x 4 infesting sporozoites
- Mature oocysts can be ingested by the HD (short direct cycle) or by intermediate hosts (indirect cycle).

- **Intermediate Host (IH):** warm-blooded animal (mammal or bird)

- Ingestion of mature oocysts containing sporozoites=> tachyzoites macrophages, blood phase of dissemination or septicaemia
- Immune response: the parasite is confined to organs with a low immune response (eye, brain, muscles)=> encysted bradyzoites waiting to develop
- Ingestion of cysts containing bradyzoites=> toxoplasmosis.

a. Asexual cycle in the intermediate host :
Human contamination is essentially dietary, either through ingestion of oocysts contained in water or food contaminated by cat faeces; or through ingestion of undercooked meat containing cysts (mutton, beef, pork, horsemeat or even poultrymeat). Congenital toxoplasmosis is caused by transplacental contamination. The asexual cycle in the intermediate host takes place in two phases:

a.1. The acute phase: In the human digestive tract, under the action of digestive juices, the oocyst or cyst will release sporozoites or bradyzoites respectively, which will spread throughout the body via the blood and lymphatic systems and transform into tachyzoites.

These invade the cells of the reticuloendothelial system and multiply, leading to their lysis and causing necrotic lesions in the parasitised tissues. This episode is dangerous for pregnant women because it exposes the foetus to transplacental passage of the parasite in the absence protective immunity during this acute phase. During this phase, the macrophages are filled with tachyzoites which have multiplied by bipartition in the parasitophorous vacuoles, forming pseudocysts so called because the host cell will burst again and release tachyzoites which will invade other cells. The host cell nucleus is pushed towards the periphery.

a.2.The chronic or latent phase: Once effective immunity has established, the parasites, once inside the cellslow down their multiplication and metabolism, transforming into bradyzoites. The parasitized cell becomes encysted, this time forming a true cyst with degeneration of the host cell nucleus. This phenomenon is mainly seen in organs with low antibody levels (brain, striated muscle and eye).

b.Cycle in definitive host :

The definitive host is represented by domestic felids (cats) and wild felids (leopards, tigers, lynxes, etc.). The definitive host hosts the complete cyclei.e. an asexual cycle and a sexual cycle. They become infected by cysts contained in their prey or even oocysts contained in water or food. The vegetative forms released in the cat's small intestine penetrate the enteroepithelial cell where they multiply, forming a schizont. This asexual multiplication is called the schizogonic phase, which lasts around 48 hours and is repeated several times. After a few schizogony, the schizozoite penetrates the enterocyte and instead of producing a schizont, it is transformed into a potentially sexual element resulting in the formation of either a male microgametocyte or a female macrogametocyte. This phase is called the gamogonic phase and constitutes the sexual cycle. The male microgametocyte will penetrate the female macrogametes and the fusion of these two sexual cells results in the formation of a non-spore-forming oocyst which is eliminated with the faeces into the environment. Sporulation occurs in the presence oxygen at a temperature below 37C. The sporulated oocyst contains 2 sporocysts, each containing 4 sporozoites. This phase is known as the sporogonic phase.

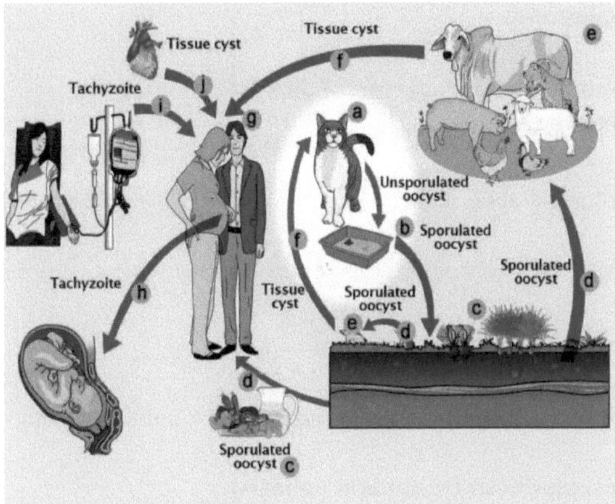

Figure 21: Evolutionary cycle of Toxoplasma gondii

Clinic

There are three clinical forms:

1. Acquired toxoplasmosis in immunocompetent individuals
2. Toxoplasmosis in immunocompromised patients.
3. Congenital toxoplasmosis

1. Acquired toxoplasmosis in immunocompetent individuals

- Latent form (asymptomatic primary infection): 80%.
- Patent form : Fever usually moderate at 38°C; Asthenia

Polyadenopathy: small lymph nodes without peri-adenitis, cervical and occipital lymph nodes persist for several weeks

Sometimes a maculopapular rash. Exceptional moderate splenomegaly
- Course: often benign.

Severe forms are exceptional

2. Toxoplasmosis in immunocompromised patients

They most often correspond to the reactivation of latent cysts with the release trophozoites (tachyzoites): HIV patients+++ or others

2.2 Cerebral toxoplasmosis :

- The most common opportunistic infection of the CNS.
- Enables the patient to be classified as having AIDS.
- Occurs late (CD4<100/mm3).
- Any central neurological abnormality in an HIV-positive patient: headache, disorientation, convulsion.
- Cerebral MRI: image single or multiple abscesses with the typical cocarde appearance.

- Favourable outcome with treatment.

2.3 Extra-cerebral toxoplasmosis: retinal (2nd location), pulmonary, etc.

2.4 Generalised toxoplasmosis (transplant patients)

It corresponds to a primary infection most often secondary to transmission via the graft during the transplantation of a solid organ from a toxoplasmosis-positive donor to a pre-transplant negative recipient. It occurs a few weeks after the transplant and immunosuppressive treatment, in a disseminated form with serious poly visceral involvement.

3. **Congenital toxoplasmosis** :

Related to the transplacental passage of tachyzoites to the foetus during a maternal primary infection. The risk of transmission is 17% in the 1st trimester, 50% in the 2nd trimester and 65% in the 3rd trimester. % 3rd trimester. In contrast, the severity decreases as the age of the pregnancy increases:

- Death in utero: encephalomyelitis (convulsions, hypertonia, macrocephaly), cerebral calcifications, microphthalmia...),

- Visceral form (anasarca, , HSPM, purpura, myocarditis, etc.)
- Late forms: psychomotor delay, pigmentary chorioretinitis.
- Latent forms: serological translation

Diagnosis of toxoplasmosis

Toxoplasmosis is diagnosed in women whose immunity is to be assessed before or during pregnancy, and in newborns where congenital toxoplasmosis is suspected. It is also requested in immunocompromised patients, particularly those with HIV and transplant patients. The diagnosis can be made using various biological, serological, histological or molecular methods.

1. Biological diagnosis

Biological diagnosis of toxoplasmosis is carried out by serology or detection of the parasite or parasitic DNA.

- Parasite testing by inoculation into mice and parasite DNA testing by PCR are recommended for the diagnosis of congenital infections and severe toxoplasmosis in immunocompromised patients. These tests can be performed on blood, bone marrow, CSF or placenta.
- Serological diagnosis is the essential element in the diagnosis of primary infection in immunocompetent patients and pregnant women.
- Serology is of little value in immunocompromised patients, but can used to identify patients at risk of reactivation (positive serology).

Prophylaxis

The aim of prevention is avoid contamination in seronegative pregnant women and immunocompromised patients (particularly because of endogenous reactivation of cysts), who are the target population. It therefore requires serological screening. In pregnant women, serological screening allows immediate treatment of toxoplasmosis acquired pregnancy.

Phylum Apicomplexa Class Sporozoea Subclass Coccidia
Suborder Haemosporina Genus Plasmodium
Species: P. falciparum; P vivax; P. malariae; P. ovale.
Malaria is a parasitosis caused by haematozoa of the genus Plasmodium, transmitted by mosquitoes of the genus Anopheles. Its distribution follows the equatorial belt and is closely linked to the distribution of its vector. To date, five species of Plasmodium are known malaria in humans. These are Plasmodium falciparum, Plasmodium vivax, Plasmodium ovale, Plasmodium malariae and Plasmodium knowlesi.

Parasite cycle

Plasmodium is a protozoan of the Apicomplexa phylum. Its development cycle is known as tenene, and requires two hosts. There is therefore a difference between the asexual phase (in HI) and the sexual phase (HV). The asexual phase: takes place humans. It is the intermediate host in the cycle, serving as a reservoir for the parasite. It takes place in several phases. Primary phase: Clinically asymptomatic. During a blood meal, the infected mosquito injects sporozoites which rapidly reach human hepatocytes. The parasite develops there over 5 to 8 days, reaching the schizont stage. When these hepatocyte schizonts rupture, they release merozoites which enter the bloodstream. In P. vivax and P. ovale some parasites can remain in the liver in dormant form, also known as hypnozoites, for weeks, months or years. Secondary phase: this marks the start of the clinically symptomatic phase. In P. falciparum, the released merozoites go on to infect the erythrocytes. The parasites develop there in an average of 48 hours, progressing through the ring (0-24 hours), trophozoite (24-36 hours) and schizont (36-48 hours) stages, until new merozoites are released. A rapid increase in parasitaemia is observed per cycle, of the order of 6 to 20 times the previous parasitaemia. At the same time, the first clinical signs appear, including fever synchronised with the bursting of the schizonts (15). Reproduction of this cycle maintains the infection (Fig.22). Sometimes certain merozoites, on infection of a new erythrocyte, take a different differentiation path leading to the production of a male or female gametocyte (9 to 11 days for P. falciparum). Gametocytes are unable to fertilise in humans. They need the anopheles to produce sexual reproduction by fertilisation of male and female gametocytes in the anopheles digestive tract.

The sexual phase: takes place in the female Anopheles, which is the definitive host. During a new blood meal, the female Anopheles will collect these gametocytes. Fertilisation results in the production of an oocyst which, after maturing and bursting, produces around 20,000 sporozoites. These sporozoites will colonise the mosquito's salivary glands and may be transmitted to humans a future blood meal.

Clinic

The clinical signs of malaria vary in expression and severity. Classically, they are divided into uncomplicated malaria attacks and complicated or severe malaria attacks. Other clinical forms include progressive visceral malaria, quartan nephritis and haemoglobinuric bilious fever.

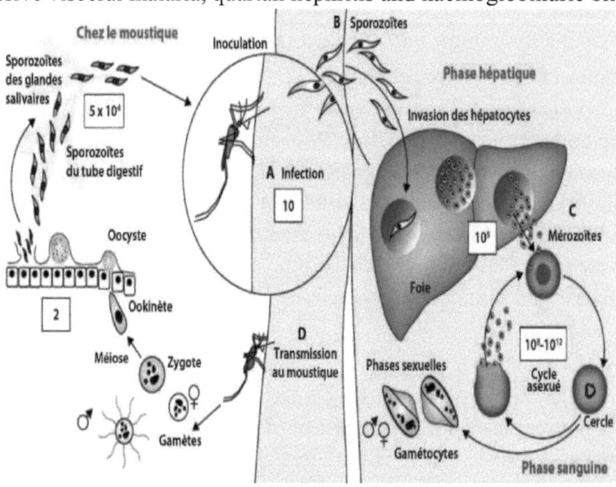

Figure 22: Plasmodium evolutionary cycle

Simple malaria access

It will be marked by a sudden onset of fever, which may be continuous or irregular, accompanied headaches, arthralgias, myalgias a digestive syndrome including anorexia, diarrhoea, nausea and vomiting, and abdominal pain. Simple malaria can develop into a severe attack at any time. The diagnosis of malaria is therefore a medical emergency.

Severe malaria or neuromalaria

"Any patient presenting an impairment of consciousness or any other sign of cerebral dysfunction on return a malaria-endemic must be treated as soon as possible.This is a major emergency such as neuromalaria. (2) Despite appropriate treatment, the Mortality from neuromalaria remains high, at up to 30%. Severe attacks often have a gradual onset, with irregular fever, an algic syndrome and digestive problems. It is most often due to infection with Plasmodium falciparum, but also Plasmodium vivax and knowlesi, and can be potentially fatal.

Evolving visceral malaria (EVM)

Evolving visceral malaria, also known as malarial cachexia or chronic malaria, is seen in hyperendemic , in children under the age of 15 who are in process of immunity, and also in people who have suffered repeated parasitic infections and are taking partially effective treatment or chemoprophylaxis. From a clinical point of view, it manifests itself by deterioration in general condition, anaemia, splenomegaly, weight loss and, in children, a delay in growth and development. Parasitaemia is low, which complicates diagnosis. However, malaria serology is positive. In the majority of cases, the disease progresses positively and rapidly after administration a chloroquine-based antimalarial treatment.

Diagnosis

Signs of orientation :

Notion of a stay in an endemic zone.
Certain neurological forms.
Changes in certain parameters: haemolytic anaemia (primary infection phase and revival phase) and thrombocytopenia (platelet count below 150,000elts/mm3) **Direct diagnosis** The diagnosis of certainty is made by detecting plasmodium in the subject's blood. The reference techniques used are still: blood smear and thick drop.

Indirect diagnosis

Serological diagnosis of malaria cannot replace direct parasitological examination, as specific antibodies appear several days after parasitaemia. Serology mainly indirect immunofluorescence, but other techniques have also been applied: ELISA, hemmaglutination. The indications for serology are :
- Screening of potentially dangerous blood and therapeutic follow-up for patients.

- In cases of fever of undetermined origin
- Epidemiological study: prevalence of malaria

Prophylaxis.

Collective prophylaxis

Vector control: Anopheles vectors is the main aspect malaria control. Various measures are recommended:

- Use of mosquito nets impregnated an insecticide and spraying insecticides
- Drainage of swamps or draining of swamps
- Biological control of Anopheles larvae (Bacillus thuringiensis, larvivorous fish, etc.) Gambusia...)

Individual prophylaxis.

In endemic , the use of Chloroquine or Amodiaquine is aimed at reducing parasitaemia, thus avoiding the fatal evolution of severe forms. In the case of travellers, the chemoprophylaxis required will take account of exposure to risk and therefore the local situation of chemoresistance.

Phylum : Ciliophora

Class: Kinetofragminophorea Order: Trichostomatida Family: Balantiididae

Genre: Balantidium

Species: Balantidium coli is specific to humans.

There are other porcine species: B. suis; B. caviae.

Balandidium coli is a cosmopolitan parasite, very common in tropical areas and especially in regions where pig is important and where pigs live more or less close contact with humans. The name Balantidium comes from the Greek word Balanto, meaning sack, due to its morphology. Balantidiosis is an intestinal protozoosis caused by a ciliate, Balandidium coli, a common parasite of pigs. It is a condition linked to faecal and water-borne perils, and mainly affects people who eat pork.

Morphology

The pathogen exists in 2 forms: the vegetative or trophozoite form and the cystic form.

a. **The vegetative or trophozoite form**

It is a large, ovoid protozoan (Fig.23). It measures 50 - 200 μm / 20 - 70μm. Its anterior end bears the cytostome (or mouth) and is more tapered than the posterior part, which opens into an anus.The granular endostome contains numerous digestive vacuoles filled with food particles (starch grains, intestinal bacteria, red blood cells) and 2 contractile vacuoles which move around cell. There are 2 nuclei: the macronucleus (a large, ovoid, kidney-shaped vegetative nucleus containing a dense chromosome) and the micronucleus (a small, rounded reproductive nucleus located in the concavity of the macronucleus). Nutrition is provided by phagocytosis of cellular debris, nutrients and intestinal bacteria. Mobility is explained by the coordinated spiral movement of its cilia or rotation. Balantidium coli multiplies in the colon in 2 particular ways: the most frequent and usual is by transverse binary division of the cell and nuclei (division of the micronucleus by mitosis, then of the macronucleus and finally of the cytoplasm and contractile vesicles). This is asexual division, and it is after a certain number of these divisions that the second type of multiplication, known as sexual multiplication (or conjugation), occurs. This involves 2 parasites coming together by partial fusion of the cytoplasmic membranes and, above all, a reciprocal exchange of chromatin material before the 2 individuals separate.

***b.* The cystic form**

It is rounded or spherical and measures 50 - 60 µm in diameter (Fig.23). Its wall is thick and transparent. The parasite, with 2 nuclei and contractile vacuoles, is mobile inside the cyst, with persistent cilia. Encystment occurs in the colon.
The cyst is the form of resistance and transmission of the parasite.

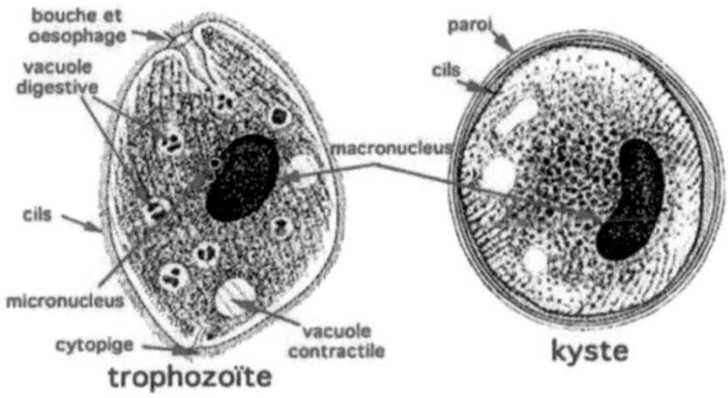

Figure 23: Diagram a vegetative form and a cyst of Balantidium coli

Habitat

Ingested cysts are found in the small intestine before they disenucleate. Trophozoites and new generations of cysts live in the colon humans and pigs. **Pathogenesis** Balantidium coli produces a proteolytic enzyme (hyaluronidase) that cuts and digests enterocytes. This causes ulceration of the colon allowing infiltration lymphocytes and polymorphonuclear cells. The lesions are ulcers in the wall of the colon, very localised and superficial. Haemorrhage occurs and a bacterial infection sets in. Perforation of the colon or appendicitis can be fatal if not treated early.

Final host

The pig is the definitive host. Man is an accidental host.

Parasite reservoirs

These are :

- From humans (80% of healthy carriers) ;
- Pork (very often parasitized throughout the world, and pig faeces are a source of parasites). infesting humans).

Mode of contamination

Transmission is **fecal-oral**, through ingestion of water or food contaminated by cysts.

Evolutionary cycle

The definitive host most often becomes infected by ingesting cysts through contaminated water or food. After ingestion, the cysts split, shed their shells and release vegetative forms in the small intestine (Fig.24). The trophozoites then colonise the large intestine. In the lumen of the large intestine, the trophozoites multiply by transverse binary division or conjugation. Some trophozoites will encyst to become infesting cysts; others will continue to colonise the colon. They may penetrate the sub-mucosa and have a lytic action on the tissues, promoting progression to other organs (broncho-alveolar and urinary tracts).
Figure 24 below shows the life cycle.

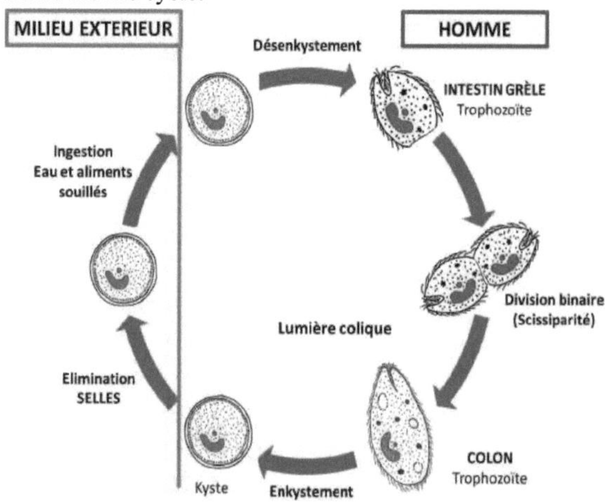

Figure 24: Life cycle of Balantidium coli

Contributing factors

- Parasite resistance: cysts survive for 15 days at a temperature 20 - 25° or 2 months at low temperatures in a humid environment;
- Occupation: farmers, pig breeders, vets, gardeners using manure ;

- Poor hygiene conditions ;
- Poor general condition of the host: malnutrition, disturbance of the bacterial flora, alcoholism, chronic gastrointestinal problems, immunodepression;
- Environment: overcrowding, tropical climate, pollution.

Clinic

Latent forms are common.

Acute forms are characterised by abdominal pain and bloody diarrhoea (simulating amoebic dysentery), which can be complicated by haemorrhage, perforation and metastatic abscesses. Hepatic abscesses and chronic balantidial colitis have been reported. Involvement of the bronchopulmonary and genital-urinary tracts with renal failure can lead to serious, difficult-to-diagnose and fatal symptoms.

Diagnosis

Elements orientation

- epidemiological data: geographical origin, profession, age, etc. ;
- the notion of travel to endemic regions /promiscuity with pigs ;

Parasitological diagnosis

Parasitological examination of the stools to detect trophozoites and cysts. In the case of complications: trophozoites may be detected in bronchoalveolar fluid or in urine during pulmonary and urinary tract infections.

Parasitological techniques for stools

- Fresh examination: Fresh direct examination enables Balantidium coli to be detected in the stool under the light microscope. This involves diluting a fragment of stool with physiological water. Observation under the light microscope reveals cysts and trophozoites, which are characterised by their rapid movement. Lugol's stain is used to visualise the double wall of the cysts.

Rectoscopy: trophozoites are sought in the lesions.

- **Concentration techniques**

They improve the sensitivity of the examination. These include Ritchie concentration and Merthiolate Iodine Formol (MIF) staining.

- Ritchie: this is a concentration technique using a formalin-ether mixture. The organic matter is dissolved in the organic phase (supernatant) while the eggs and cysts of the intestinal parasites are found in the aqueous phase (pellet). The pellet containing the parasites is observed after centrifugation between slide and coverslip.
- MIF: The Merthiolate - Iodine - Formol mixture is used to stain cysts and vegetative

forms; it is also used to preserve stools to prevent any deterioration of the cysts. The use of MIF has the advantage of allowing delayed examinations.

Specific immunological diagnosis using the indirect immunofluorescence technique. It is used to measure anti-Balantidium antibodies.

Molecular diagnosis using the PCR technique.

Prophylaxis Individual prophylaxis

- Hygiene of hands, food and water.

Collective prophylaxis

- Avoid human and pig promiscuity;
- water treatment
- Construction of latrines ;
- Therapeutic management of asymptomatic carriers in zones endemic areas ;
- Health education for populations in endemic areas ;
- Monitoring pigs

III. PARASITIC METAZOA

1- Phylum of Plathelminthes

Class Trematoda Subclass Digenea Genus Fasciola
Genus Paragonimus

Definition

Distomatoses are anthropozoonoses caused by the presence in the body of hermaphrodite trematodes: flukes or distomes. Distomatoses are favoured by non-compliance with food hygiene rules and low socio-economic status. The flukes responsible for distomatosis belong to several orders and various families of trematodes. They are digenes, i.e. they have two distinct modes of reproduction during their life cycle: sexual, with the production of eggs, and asexual, with multiplication at the larval stage in an intermediate mollusc host.

Morphology

Adults

Staves are flat, non-segmented worms with a foliaceous (leaf-shaped) body, ranging in size from 1mm (Heterophyes heterophyes) to 8cm (Fasciola gigantica). They are equipped 2 suction cups enabling them to attach themselves to the bile ducts, intestine or lungs, depending on the species (Fig. 25). The mouth opens at the centre of the anterior or oral sucker. The ventral sucker is called the acetabulum. Flukes hermaphroditic parasites that feed on liquid or near-liquid substances taken their living environment. These substances are often totally or partially digested. The hole in the centre of the anterior sucker serves as both the animal's mouth and anus. Digenetic trematodes living in the digestive tract of the final host are essentially anaerobic. The adults are hermaphroditic and lay non-embryonated eggs, except in the case of Dicrocoelium dendriticum.

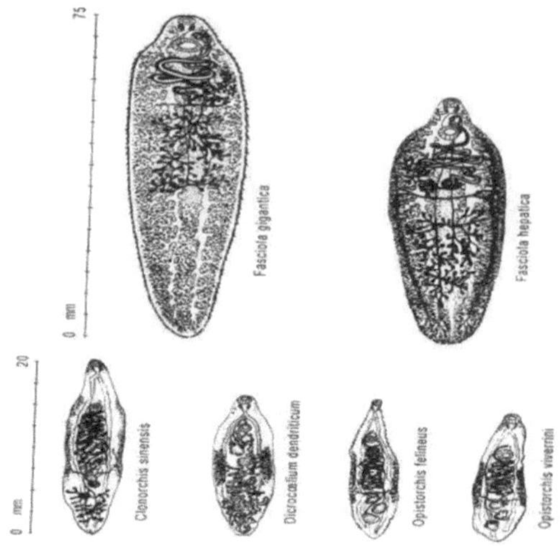

Figure 25: Adult flukes causing hepatic distomatosis (Ripert C, 1990)

Eggs

The eggs come in a variety of shapes but are generally ovoid, measuring between 20 and 200µm in size. diameter. These are capped elements that are not embryonated at oviposition, except for those of D. dendriticum.

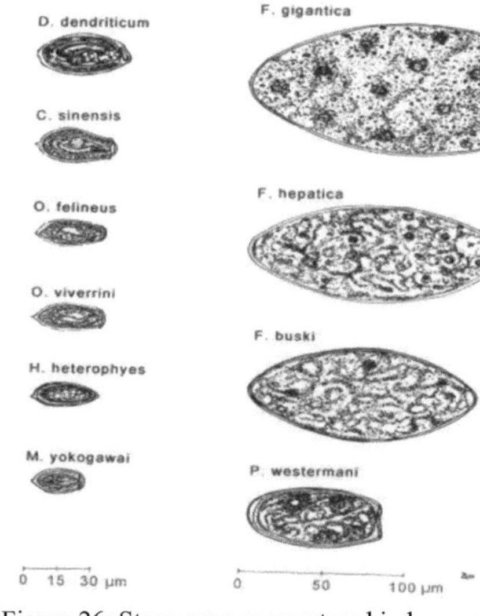

Figure 26: Stave eggs encountered in human parasitology

Pathogenesis

Distomata parasitising humans live in the bile ducts, intestines or lungs. The secretions and excretions of the worms lead to sensitisation reactions and a humoral immune response. Chronic irritation of the epithelium of the bile ducts and bronchi leads to oedema, an inflammatory reaction and reactive hyperplasia. The habitat of adults enables them to be classified according to the damage caused to these different organs in the definitive host:

a. Hepatobiliary distomatosis :

Fasciola hepatica Fasciola gigantica Dicrocoelium dendriticum Clonorchis sinensis Opistorchis felineus Opisthorchis viverrini

b. Intestinal distomatoses :

Fasciolopsis buski Metagonimus yokogawai Heterophyes heterophyes

c. Pulmonary distomatoses: Paragonimus westermani Paragonimus kellicoti Paragonimus africanus

Final host

Distomatoses are anthropozoonoses for which man is an accidental host. Vertebrates (cattle, sheep, etc) are the definitive host of these parasites. The domestic and wild cattle, sheep and goats, dogs, cats, pigs and other wild mammals.

Intermediate host

Each fluke species has 2 intermediate hosts (HI): a mollusc and an animal or plant. The first intermediate host is a gastropod mollusc, while the second is either a plant, a fish, a crustacean or an ant. These last four hosts are also known as **encystment hosts.**

The aquatic intermediate hosts of distomes that are parasitic on humans are found only in freshwater. They belong to several genera, including Bithynia, Galba, Melania and Zebrina.

Mode of contamination

People are contaminated **by** ingesting raw or undercooked plants or animals, containing live fluke metacercariae.

Output channel

Parasite eggs are eliminated in the faeces in the case of intestinal and hepatobiliary distomatosis. They are eliminated in the sputum for pulmonary distomatosis. Pulmonary distomatosis eggs can be found in the faeces after swallowing sputum containing them.

Life cycle

The adult fluke lays eggs characterised by the presence an operculum. The eggs are expelled with the faeces (hepatobiliary or intestinal distomatosis) or sputum (pulmonary distomatosis). With the exception of D. dentriticum, an aquatic environment is necessary for the evolutionary cycle to continue. In an aquatic environment, the egg embryonates in 3 weeks and releases the ciliated larva, the miracidium (Larva1). The mobile miracidium swims in search of its 1st HI, a gastropod mollusc. The life of the miracidium is short (a few hours), if it does not find its specific mollusc, it dies. When the miracidium encounters its intermediate host, it breaks into the mollusc's body and migrates to the mollusc's pulmonary cavity, where it actively multiplies by polyembryony, producing thousands sporocysts (Larva2). Each sporocyst buds and gives rise to a 3rd larval form: the redia, which is an elongated organism with an outline of a digestive tract. The redia reaches the mollusc's hepatopancreas and at this point, each redia gives rise a 4th larval form: the cercaria with a simple, non-forked tail. The cercariae break out of the mollusc's body and swim in search of their 2nd intermediate host (a plant or animal). On contact with the second HI, the cercariae attaches itself to it (in the case of plants) or penetrates it (in the case of animals, for example crustaceans in the case of Paragonimus westermani), loses its tail and surrounds itself with a thick wall and encysts in the form of a metacercariae (Larva 5), the infesting form of the fluke. In the case of D. dendricutum, the cercariae engulfed in mucus is swallowed by the ant where it encysts in the sub-oesophageal ganglion. Man is contaminated by ingestion these live metacercariae with plants (Fig.28) or raw or undercooked HI animals (Fig.27). After ingestion, the digestive juices lyse the shell, releasing the dodule, which then spreads to the rest of the body. the organ in which it reaches the adult stage.

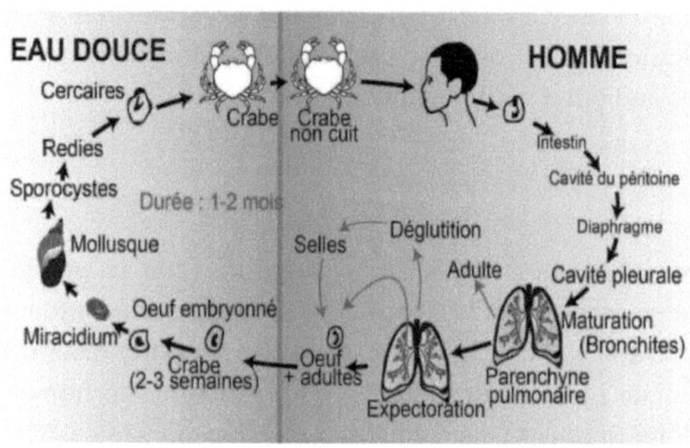

Figure 27: Evolutionary cycle of Paragonimus sp

Geographical breakdown

Distomatoses are cosmopolitan parasites. However, the species are distributed differently around the world. Fasciola hepatica and Dicrocoelium dendriticum are cosmopolitan anthropozoonoses found in all livestock regions.

Fasciola gigantica is mainly found in tropical areas.

Species of the genus Paragonimus are found in Africa and Asia.

Intestinal distomatosis is more common in South-East Asia and Tunisia.

Clinic

A.Hepato-biliary distomatosis or Fasciolosis

- Silent incubation lasting around 15 .

-In invasion or toxi-infectious phase, the stool creates lesions as it travels through the liver parenchyma. Pain is felt in the right hypochondrium, hepatomegaly, asthenia, digestive disorders (diarrhoea or constipation), pruritus, night-time fever, arthralgias, myalgias and change in general condition.

- In the state phase or chronic angiocholitis phase, the adult staves are located in the lumen of the bile ducts, which are then dilated with a thick, hard wall due inflammatory reactions of the submucosa. These flukes constitute at least a partial obstacle to the flow of bile. They may release toxic allergenic substances which cause distant skin, respiratory or nervous disorders. During this phaseasthenia and weight loss persist, as do attacks urticaria, migraines, hepatic colic, episodes jaundice with food intolerance and, more rarely, pulmonary and nervous symptoms. The disease progresses to cirrhosis.

Diagnosis

Biological diagnosis

- High blood hypereosinophilia in invasion phase and moderate hypereosinophilia in the invasion phase. state.
- An inflammatory syndrome (elevated sedimentation rate and C-reactive protein).
- Hepatic cytolysis: elevation of transaminases.

Diagnosis Parasitological examination of stools

Freshly passed faeces will be subjected to direct microscopic examination. Given the high frequency of pauciparasitism in humans, it is appropriate to use a standard method of concentration such as the Ritchie technique. The egg is large, measuring 120-140µm, light brown, non-embryonated and capped.

Histological examination

Adults may be discovered during surgical exploration of the bile ducts.

Specific immunological diagnosis

Serology is very useful for screening during the invasion phase and also for monitoring treatment. Diagnosis is confirmed by the detection of specific antibodies. Various techniques can be used: IFI, HAI, ELISA.

Radiographic examination

Liver ultrasound and CT scan can detect irregular hypodense areas in the liver parenchyma.

Prophylaxis

Combating faecal peril.

Avoid eating unwashed or badly washed aquatic plants in endemic areas. Control of intermediate host molluscs by cleaning up the environment, draining flooded meadows, rectifying banks and ditches, cleaning out drains and draining shallows, which are breeding grounds for snails.

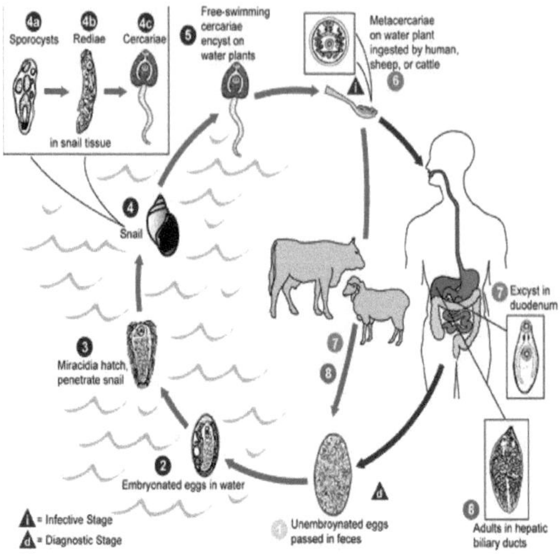

Figure 28: Evolutionary cycle of Fasciola hepatica

Kingdom: animal
Phylum : Plathelminthes
Class: Cestodes
Subclass: Eucestoda **Order:** Cyclophyllides **Family:** Taeniidae **Genus:** Tœnia
Species: T. saginata, T. solium

Taeniasis is a benign, cosmopolitan parasitic disease caused by the presence adult segmented flatworms (plathelminthes) or cestodes in the human digestive tract. Three species of tapeworm cause taeniasis in humans: Taenia solium, Taenia saginata and Taenia asiatica (identified in 1980 in Taiwan in the liver and viscera of pigs)1. However, in this chapter on taeniasis, we will focus on just two species: Taenia saginata and T. solium.

1. **Pathogens**

They are large flat worms (4 to 10 m long for T. saginata and 1 to 8 m for T. solium), ribboned and segmented into rings or proglottids (Fig.29). Following the head or **scolex**, the neck a few mm long and gives rise to the rings or **proglottids**. Together, the proglottids form the body of the tapeworm or **strobile**. The proglottids near the neck are wider than they are long, small and undifferentiated. When mature, they take on the shape of a "flat noodle"; they are rectangular, longer than they are wide (16 to 20 mm x 5 to 7 mm).

Taenia saginata

The head or scolex is pyriform, depressed at the top, with a diameter of 1.5 to 2mm in T. saginata The scolex has 4 elliptical suckers 0.7 to 0.8mm in diameter(Fig.30a), but has neither rostrum nor hook, hence its name of inerme tapeworm. At maturity, the adult measures 10-

12m long.

Figure 29: General morphology of Taenia saginata on the left and T. solium on the right

Taenia solium

The scolex is spherical, 1 mm in diameter in T. solium. It also has 4 rounded suckers (Fig.30b). and a short rostrum fitted a double ring of hooks, hence its name armed tapeworm. There are 22 to 32 hooks, with alternating large hooks (160 to 180µm) and small hooks (110 to 140µm). Less long than T. saginata, it measures 6-8 metres.

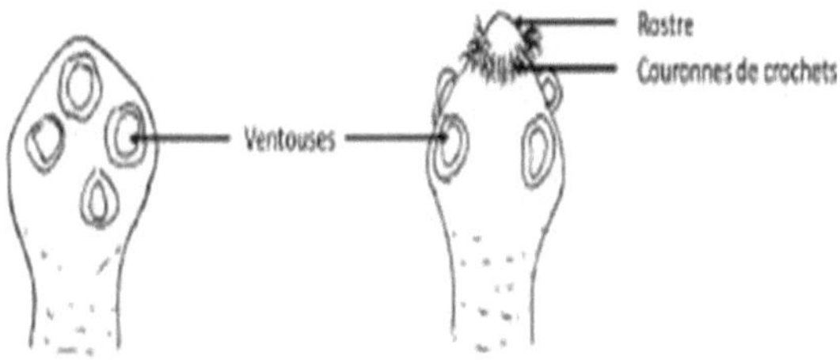

Figure 30: Scolex of Taenia saginata on the left and T. solium on the right

From one proglottid to the next, the genital pores are irregularly alternated in T. saginata and regularly alternated in T. solium (Fig.31). The reproductive system consists of 2 lobes ovaries, a cluster of 300 to 1,200 testicular glands and a highly developed, branched uterus (more than 15 dichotomous branches on each side of the central axis) filled eggs in T. saginata and the uterine ramifications are fewer (7 to 13 on each side the central axis of the proglottids) and dendritic in T. solium. Proglottids are emitted either individually by forcing the anal sphincter, outside the faeces (T. saginata), passively in more or less long chains of 10 rings (T. solium) in the external environment through the faeces. This mode of elimination of the rings, which characterises one species or the other, determines the diagnostic technique used in the testing of these worms. The adult T. saginata lives attached to the duodenal mucosa by its scolex, whereas the adult T. solium lives in the small intestine.

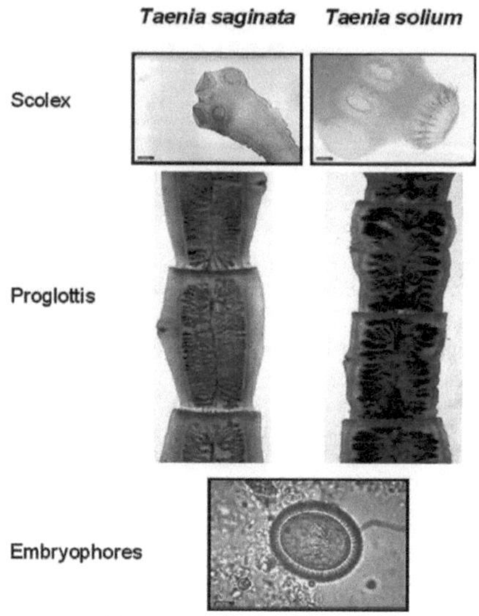

Figure 31: Parasitic elements of T. saginata and T. solium

2. Definitive host (HD)

Man is the only definitive host (HD) of Taenia saginata and T. solium. It hosts the adult of the sexually reproducing worm.

3. Intermediate hosts (HI)

The intermediate hosts (IH) of T. saginata are cattle (oxen, buffalo, etc.), whereas Suidae (pigs) are the hosts of T. solium. However, in addition to playing the role of HD, humans can also accidentally become HIs for T. solium through the dissemination of larvae in various organs, resulting in subcutaneous, muscular, neurological or ocular **cysticercosis**, etc.

4. Mode of contamination

Human infection (HD) occurs via the oral route through ingestion of T. saginata and/or T. solium larvae contained in raw or undercooked beef or pork (intermediate hosts).

6. Life cycle

In T. saginata, the rings detach **individually from** the chain and **actively** leave the intestine by forcing the anal sphincter, ending up in linen or bedding. After drying, the rings rupture to release the embryophores.

In T. solium, on the other hand, the rings **are passively** eliminated in **groups of 4-10** in the faeces. When the rings are destroyed, thousands of **embryophores** are released into the environment (Fig.32). Once the embryophores are ingested by herbivores (Bovidae, Suidae), **the hexacanth embryo** (six hooks) released in the stomach perforates the digestive tract and invades the entire organism. Once it reaches the muscles, it attaches itself, vesiculates and buds. The protoscolex (scolex of the future tapeworm) differentiates inside the vesicle and remains invaginated. The cysticerci larva (infesting form of the parasite) matures after 3-4 months.

People become infected by ingesting raw or undercooked beef or pork containing **cysticerci larvae** (T. saginata or T. solium). In the stomach, the scolex evaginates and migrates. into the intestine, where it attaches itself to the mucosa. It becomes an adult after 2-3 months and begins to eliminate its rings with the faeces (Fig.32).

In T. solium, contamination occurs either **exogenously** by ingestion of the embryophores, or **endogenously** (self-infestation). In the latter case, the rings are regurgitated into the stomach where they are digested. The hexacanth embryos are released, pass into the circulation and disperse throughout the body (muscles, brain, eycs, skin, etc.) where they encyst in the form of cysticercarial larvae. **This is cysticercosis.**

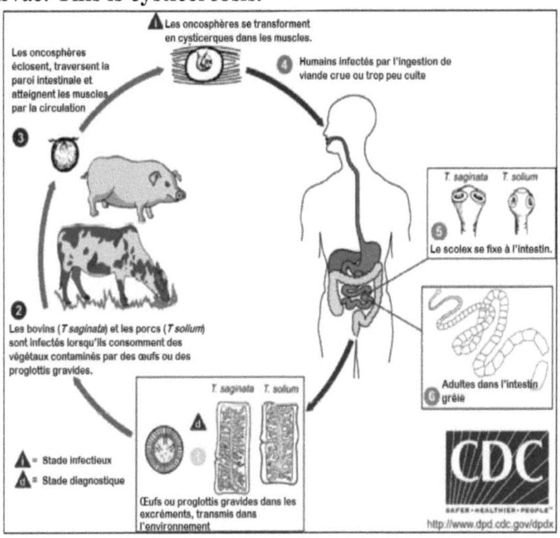

Figure 32. Evolutionary cycle of Taenia saginata and T. solium

9. Geographical breakdown

The prevalence of taeniasis is very high and depends on the level of local hygiene, the quality of veterinary health controls and people's eating habits: consumption of rare or very well-cooked meat, ban on pork. Taenia saginata is found throughout the country where is consumed. Its distribution is rare in Muslim and Jewish communities, but common in the rest of the continent.

Clinic

Invasion phase: Six to eight weeks after ingestion, symptoms may appear in the digestive tract, such as nausea, diarrhoea or constipation and abdominal pain. Biological signs such as hypereosinophilia mark this phase.

State : Clinical manifestations may occur in the form of digestive signs, but many patients remain asymptomatic:

- Bulimia or anorexia, nausea or vomiting, transit disorders alternating diarrhoea and constipation.
- Epigastric or pseudo-appendicular pain.
- **Anal pruritus.**
- Macrocytic megaloblastic anaemia is most often a sign of teniasis.
- , a tapeworm can live and produce proglottids for decades.
- There may be rare complications in T. saginata taeniasis: acute or chronic appendicitis, intestinal obstruction, perforation.

Cysticercosis caused by Cysticercus cellulosae

Only T. solium causes serious health problems because of the spread of its larvae (Cysticercus cellulosae) in human tissue, which becomes an intermediate host. People contract taeniasis by ingesting raw or undercooked meat containing tapeworm larvae. When T. solium larvae penetrate the central nervous system, they can cause neurological symptoms, including epileptic seizures. T. solium is responsible for 30% of epilepsy cases in many endemic areas, where there is close proximity between local people and free-ranging pigs.

Diagnosis

Parasitological examination of stools

It is based on the identification of the rings and the search for eggs or embryophores in the faeces (T. solium) or on the anal margin (T. saginata).

Note: The embryophores of T. saginata and T. solium are rarely observed conventional parasitological examination of faeces. They are more frequently found on anal margin, particularly in the case of T. saginata, whose rings are actively eliminated by forcing the anal sphincter.

In this case, **Graham's adhesive cellophane** test would be the appropriate technique for diagnosis. This test consists of applying a piece of transparent adhesive tape to the anal margin; the tape is then stuck to a glass slide observed under a microscope. This test should be carried out in the morning before the patient is washed. It reveals the embryophores of T. saginata, and rarely those of T. solium. In practice, T. saginata and T. solium embryophores are very difficult to differentiate.

Taenia antigen testing: the coproantigen enzyme-linked immunosorbent assay (CoAg-ELISA) is used to detect specific tapeworm antigens in stools. The CoAg-ELISA has better diagnostic sensitivity for T. solium and is currently the most reliable test for diagnosing taeniasis. This ELISA is performed using hyperimmune rabbit sera raised against somatic antigens of the adult T. solium worm. In In epidemiological studies, this test detects around 2.5 times more cases of taeniasis than microscopy.

Molecular diagnosis: PCR

Prophylaxis

Individual measures

- Avoid eating raw or undercooked meat (beef, pork, fish).

- Treatment of parasites ;
- Low and extended freezing (meat, fish).

Collective measures
- Screening for cysticerci at slaughter in HI ;
- Health education for the population (hygiene, not fertilising the soil with untreated human faeces);
- Waste water treatment.

The fight against faecal peril through the construction of latrines, the control of livestock farms with anti-parasite treatment of animals, the reinforcement of sanitary controls, massive administration of drugs against tapeworms: treating infected subjects, not using untreated human faeces to fertilise agricultural areas.

Kingdom : Animal

Phylum: Plathelminthes Class: Cestodes
Order :Pseudophyllidae Family :Diphyllobothridae Genus : Diphyllobothrium. Species :D. latum

Bothriocephalosis, also known as diphyllobothriosis, is a zoonotic parasitic disease caused by a flatworm of the genus Diphyllobothrium spp, transmitted to humans by ingestion of raw or undercooked fish. In this chapter, we will only discuss Diphyllobothrium latum diphyllobothriosis, which is the most widespread and species. The adult worm appears to be the largest of the human tapeworms. The body is divided into three parts: the head, the neck and the body. The head or tiny scolex (1 to 5mm) has no rostrum or hooks and is oblong with two longitudinal slit-like attachment organs (bothridia). The neck is a narrow germinative zone following on from the head. The body or strobile is fairly long (10 to 12 m) and is formed by a ribbon-like chain of successive segments (3,000-4,000) that are wider than they are long and trapezoidal in appearance. The segments or proglottids are hermaphroditic and the genital pores are median and ventral. It lives in the small intestine of the host (Fig.33).

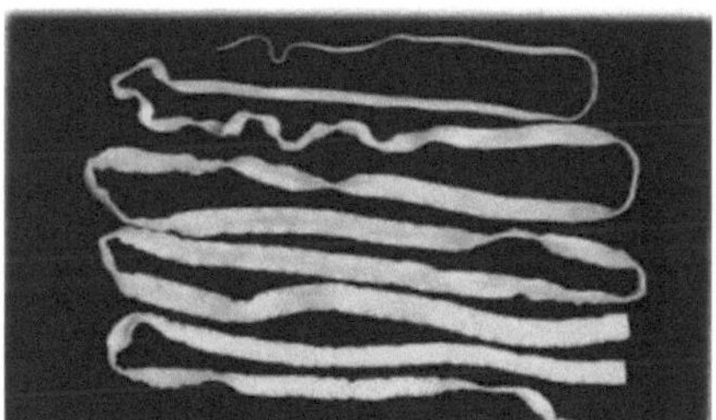

Figure 33. General morphology of Diphyllobothrium latum

Egg: 70/60 µm long, capped and not embryonated when laid; it is eliminated with the faeces into the outside environment (fresh water) where it matures over several days to weeks (Fig.34).

Coracidium (Larva 1): this is a round, swimming larva covered with vibratile cilia which comes from hatching of the egg after lifting its lid (Fig.34).

Procercoid larva (Larva 2): this is the second larval form resulting from the development of the coracidium inside a small copepod crustacean, the cyclops.

Plerocercoid larva (Larva 3): this results from the transformation of the procercoid larva to the plerocercoid larva. inside the fish that has ingested the cyclops. This is the infesting form of the parasite.

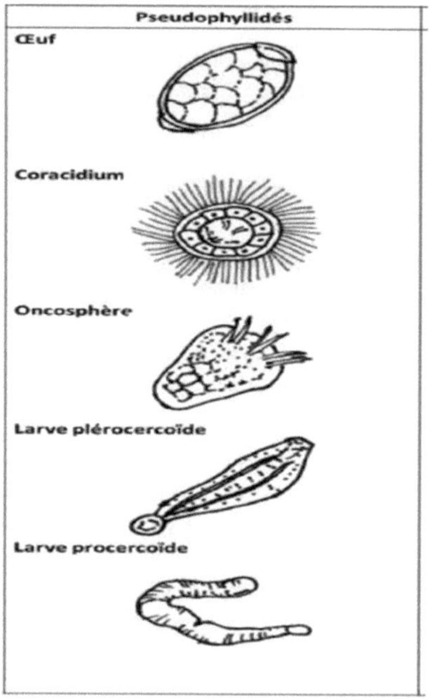

Figure 34: Parasitic elements of Pseudophyllides (Diphyllobothrium latum)

2. Final host

Humans, dogs, cats and itchiophagous animals (seals, bears).

3. Intermediate hosts

There are two intermediate hosts, a freshwater crustacean (cyclops) and a freshwater fish.

4. Mode of contamination

Contamination occurs orally through the ingestion of raw or badly cooked infested fish.

6. Life cycle

The Diphylobothrium latum cycle involves two intermediate hosts (a copepod crustacean and a fish), a paratenic host (a fish that harbours the infesting form of the parasite) and a definitive host (generally a fish-eating mammal, including humans), which harbours the adult parasite. The cycle begins when the non-embryonated eggs or oncospheres present in the faeces of the mammalian host are released into the water (Fig. 35). These hatch into a ciliated larva called Coracidium (Larva1) after 3 weeks. After being ingested by a freshwater crustacean, such a Cyclops (the first intermediate host), the Coracidium transforms into a procercoid larva (Larva 2) in 15 days. If the crustacean is in turn ingested by a second intermediate host, usually another small freshwater fish, the procercoid larva encapsulates and develops into a plerocercoid larva (Larva 3). The plerocercoid larvae constitute the infectious form of the parasite for the definitive host, including humans. Contamination of mammalian hosts begins with the consumption small raw, undercooked or smoked fish. The continuity of the cycle is further encouraged by the fact that small fish can in turn become prey for large predatory fish. By ingesting infested small fish, these large fish also accumulate their parasites. They are therefore referred to as "paratenic hosts" or "holding hosts", whose main characteristics are as follows:

- They form part of the food chain of the definitive hosts.
- They contribute to the dissemination of the parasite and constitute its reservoir.
- They are located between the definitive host and the parasite.

After being ingested by the definitive hosts, the plerocercoid larvae continue their development to become mature adult tapeworms in the small intestine in about a month. In the definitive host, the adults attach themselves to the intestinal mucosa using two bilateral hooks called bothridia. Tapeworms can live for up to 30 years, reaching lengths of more than 10 metres (sometimes up to 30 metres) and possessing more than 3,000 proglottids. After fertilisation, immature eggs are released from the proglottids (up to 1,000,000 eggs per day per worm) and excreted in the faeces. The eggs appear in the faeces around 5 to 6 weeks after the start of the infection. Their development is favoured by temperatures between 10 and 20°C. Bothriocephalosis is a parasitic disease associated with faecal peril, where eating habits (such as eating raw or undercooked fish) and environmental conditions are the main factors favouring transmission of the parasite.

Clinic

Bothriocephalosis is a generally asymptomatic disease. Those affected are often alerted by the appearance of segments in their stools. Non-specific clinical signs include abdominal pain, transit disorders, diarrhoea, fatigue, constipation, dizziness, numbness in the extremities, feelings of hunger and headaches. Complications may arise in event of massive infestations: intestinal obstructions, cholecystitis or cholangitis. Anaemia may also occur (due to vitamin B12 deficiency).

Biological diagnosis

Parasitological examination of stools: This is based on the detection of eggs under the microscope. or proglottids in the stools.

Molecular diagnosis: Molecular analysis by PCR allows precise identification of the species.

Prophylaxis

- Avoid eating raw, undercooked or smoked fish.
- Freeze fish to be eaten raw for at least a week.
- Antihelminthic treatment.
-Prevent water contamination by encouraging the use of latrines and septic tanks for excrement management.

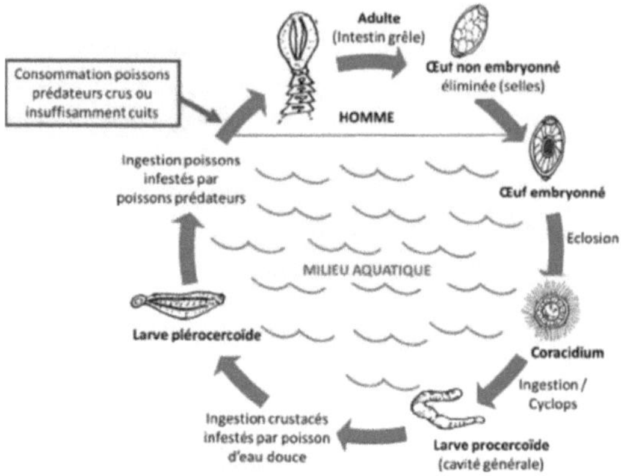

Figure 35: Life cycle of Diphyllobothrium latum

Phylum: Platheminthes Class: Cestodes

Order: Cyclophyllida Family: Taenidae Genus: Echinococcus
Species: Echinococcus granulosus

Hydatidosis (also known as hydatid echinococcosis or hydatid cyst) is a zoonosis caused by the development in humans of the hydatid larva of the tapeworm Echinococcus granulosus. It is transmitted to humans ingestion of the eggs (embryophores) excreted into the environment by the parasitised dog. This condition is widespread in farming areas. The adult worm is the smallest cestode that lives in the digestive tract of dogs and certain canids. **The adult** is the smallest tapeworm, measuring 3 to 8 mm (Fig.36). The scolex has 4 rounded suckers with a rostrum armed with a double ring of hooks. The body has 3 to 4 rings, only the last ring is gravid and contains a uterus filled several hundred embryophores. The penultimate ring, which is longer than half the body, has a developed hermaphroditic genitalia, with an ovary and 20 to 70 testes.

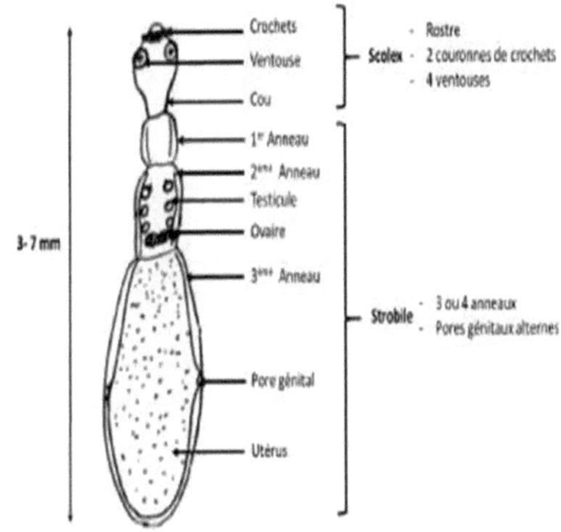

The hydatid larva: this is the largest larva and is hollow, filled with liquid, vesicles and scolex (protoscolex). The fertile hydatid includes (Fig.37) :

- A thick, stratified outer cuticular membrane, 0.5 to 1 mm thick, which has the consistency and colour of cooked egg white. This membrane is made up of lipids, proteins and mucopolysaccharides, and is made up of concentric layers nested one on top of the other;

- An internal germinative membrane or proligeral membrane 10 to 25µm thick containing embryonic cells. Within the hydatid, this gives rise to 250-500µm proligeral vesicles by budding. Inside these vesicles, the scolexes bud;

- Hydatid fluid fills the hydatid and vesicles. It contains numerous protoscolex and membrane debris known as hydatid sand. There are also exogenous daughter vesicles that are expelled from hydatid.

These daughter vesicles have the same structure as the mother hydatid. A hydatid cyst is a collection of hydatid larvae surrounded by a fibrous reaction. of the host, known as the adventitious membrane.

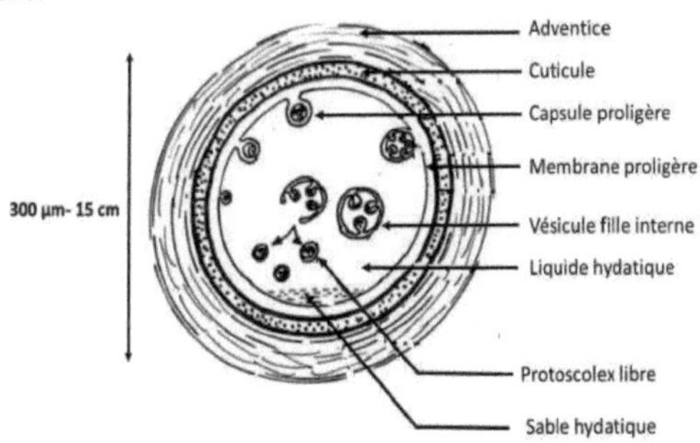

Fig.37 : Structure of the hydatid cyst

The embryophore

Slightly oval in shape, its morphology is similar to that of the eggs of T. saginata and T. solium. dimensions from 35 to 45μm

c. Biology

The adults usually live in the dog's intestine, which harbours a large number of them without being particularly bothered. The vesicular larva or hydatid cyst is found in 50 to 60% of cases in the liver, in 30 to 40% of cases in the lung and in 5 to 10% of cases in any other part of body of the infected individual (heart, brain, spleen, kidney, skeleton, muscles). The embryophores are released with dog faeces and can survive in the environment for 18 months to 2 years, at temperatures ranging from +25°C to -25°C.

2. Final host

The dog and certain wild canids (wolf, jackal, fox).

3. Intermediate hosts

Herbivorous or omnivorous mammals, including sheep and occasionally humans, are the main intermediate hosts of E. granulosus.

4. Mode of contamination

Humans become infected by ingesting the embryonated eggs (embryophores) eliminated into the external environment by the definitive host, which may be present on dogs or on soiled soil and foodstuffs. Infection of humans constitutes an evolutionary parasitic dead end for the larva, which can no longer continue its development.

Definitive hosts become infected by eating the hydatid viscera of a parasitised intermediate host (sheep+++).

Parasite reservoir

These are the canids that host the adult form. The larvae are found in the hosts intermediates that ensure contamination of the definitive host by carnivory.

7. Natural" evolutionary cycle

E. granulosus embryophores are released into the environment with the faeces of the parasitised dog. Sheep (or other herbivores) become infected by grazing on the contaminated grass (Fig.38). The egg hatches in its digestive tract, releasing a hexacanth embryo which reaches the liver or, more rarely, other organs where it transforms into a hydatid larva. The dog (definitive host) then infests itself by devouring the hydatid viscera of the parasitised sheep. In dog's intestine, the scolex disinvaginates and attaches itself to the mucosa, giving rise to an adult in 6 to 8 weeks. From then on, it begins to excrete eggs (embryophores) with the faeces into the environment. The embryophores can survive in the outside environment for 18 months to 2 years, at temperatures varying between 0°C and 25°C.

□ **Accidental cycle in humans**

Humans are accidental intermediate hosts, becoming infected by ingesting embryophores. eliminated by the faeces of the parasitized dog (Fig.38). The wall of the embryophore is dissolved in the stomach and the hexacanth embryo is released. Thanks to its hooks, it crosses the wall of the small intestine, reaching the liver where it settles in 50 to 60% of cases; it can be found in the lung (30 to 40% of cases) other organs such as the brain, heart, kidneys and spleen (5 to 10%). It vesiculates from the 7th day, surrounds itself with an inflammatory granuloma and gives rise to a hydatid larva which matures in 12 to 18 months.

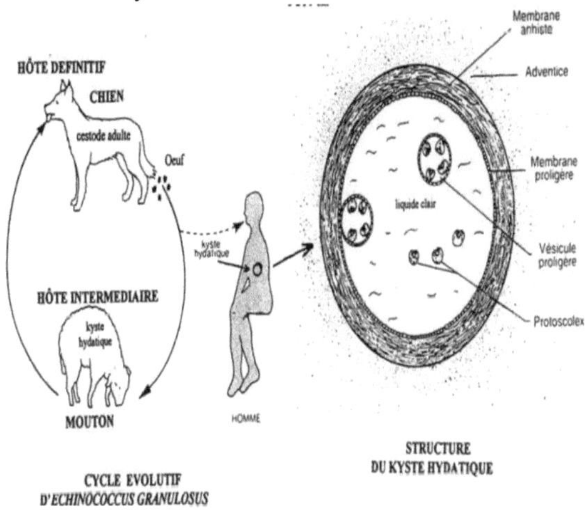

Figure 38: Evolutionary cycle of Echinococcus granulosus

8. Factors favouring the transmission of hydatidosis

- Hydatidosis is more common in farming areas, particularly where large flocks of sheep are guarded by dogs.
- Disposal of animal carcasses in the wild, uncontrolled slaughter of animals.
- No antihelminthic treatment for dogs.
- Profession: breeders, shepherds, hunters in close contact with dogs.
- Eating habits: consumption raw or undercooked food.
- Faulty hygiene.

Clinical :

In liver disease: more or less painful hepatomegaly or sometimes a rounded tumour bulging under the costal crest, which may be painless or tender and well limited. Compression of the bile ducts with jaundice, hepatic colic and sometimes angiocholitis in the case of fissuring. Pulmonary involvement: haemoptysis, dyspnoea, lung abscess with vomit consisting a more or less clear liquid containing hooks and scolex.

Diagnosis FNS

Hyper eosinophilia in the blood during the parasite growth phase or in the event of rupture or fissure.An increase in blood bilirubin levels.

Neutrophil hyperleukocytosis may be present in cases of bacterial superinfection.

Biological diagnosis

Imaging tests to identify the hydatid cyst: abdominal ultrasound, chest X-ray.

Immunological diagnosis

Immunological diagnosis is based on the detection of specific serum antibodies using a variety of techniques, including qualitative techniques such as immunoelectrophoresis and electrosyneresis, and quantitative techniques such as indirect immunofluorescence, ELISA and haemagglutination.

Serology is of interest in the therapeutic monitoring of the disease.

Prophylaxis

1. Individual prevention

- Reduced promiscuity with dogs.
- Good personal hygiene: wash your hands after contact with a dog.
- Food hygiene: washing food, especially when there are dogs around.
- water consumption.

2. Collective prevention

- Deworming of dogs, elimination of stray dogs.
- Control animal slaughter using official abattoirs.
- Destruction of the viscera of infested animals.
- Surveillance of people at risk: farmers, shepherds, hunters.

Phylum of the Nematoda,

□ Class Secernanta,□ Order Ascaridida,□ Family Oxyuridae,□ Genus Enterobius,

□ Species: Enterobius vermicularis Linné 1758.

Oxyurosis, also known as Enterobiasis, is an intestinal helminthiasis caused by the development of a nematode in the human small intestine: Enterobius vermicularis or pinworm. It is a cosmopolitan disease that is widespread throughout the world, affecting children in particular.

2. Morphology

a.The adult worm

It is a round, white worm, visible to the naked eye. The male measures 2 to 5 mm in length and 0.1 to 0.2 mm in diameter. Its posterior end is curved ventrally and bears a copulatory spicule which is the terminal end a genital apparatus consisting a testis, a seminal vesicle and

an ejaculatory duct.
The female measures 9 to 13 mm in length and 0.3 to 0.5 mm in diameter. Its body is straight or slightly arched with a thicker middle section. The posterior end is straight, tapering and pointed. The uterus is filled with thousands of eggs, causing it to distend. The vulva opens in the anterior third of the body.
The cephalic part bears the mouth, which is surrounded by 3 retractile lips. It has a cuticular, vesicular and transversely striated swelling. The oesophagus is muscular with a characteristic posterior bulb.

b. Egg

It is oval and asymmetrical, with one convex side and one flat side. It measures 50 to 60 µm by 30µm. It has a thick, smooth, transparent and refractive shell. The outer surface is albuminous, allowing the egg to adhere to the skin or mucosa. It contains a vermiform embryo.

B. Final host

Humans are the parasite's definitive host.

C. Methods of contamination□ by Oral route:

- Or by inter-contamination: directly from dirty hands. This ensures inter-human dissemination. Indirectly via soiled objects (sheets, toys), water and food;
- Or by self-contamination: the eggs laid adhere to the anal margin and stick to the skin. and the hands brought to the mouth giving a new infestation.

F. Evolutionary cycle

Humans become infected by ingesting eggs through dirty hands or soiled food (Fig.39). In the stomach, these eggs hatch into rhabditoid larvae which enter the small intestine. After three moults, these larvae become adults in 15 to 40 days. In the early stages of their lives, young adults live in the ileum and feed on food debris. They migrate to the cœco-appendicular region where mating takes place. The pregnant female migrates through the colon to the rectum during the night. She actively crosses the anal sphincter. She bites the anal mucosa to hold on and starts laying eggs as soon as she comes out into the open. They lay around 10,000 eggs in 20 minutes, after which they usually die. In girls, they can migrate to the genital tract, leading to genital clinical manifestations. The eggs are not embryonated at the slope. They become mature and infective in 6 hours at a temperature of 37°C in the presence of oxygen. The average cycle lasts between 15 and 45 days. The eggs live in the wild for several weeks depending on climatic conditions (3 weeks between 17 and 20°C. They can withstand desiccation for around ten days).

G. Contributing factors

- Grouping of people (crèches, boarding schools, hospitals, etc.)
- Poverty, socio-economic level and poor hygiene.

Clinic

It is a benign disease that usually goes unnoticed when the infestation first occurs.
In event of massive infestation, the following can be observed:
- Anal pruritus, especially at night.
- Digestive signs: pain, diarrhoea, nausea, vomiting, anorexia.
- Behavioural problems: irritability, aggression, restlessness at night.
- Vulvovaginitis with vulvar pruritus and leucorrhoea in the girl.
- Anal dermatitis is rare but possible.

Complications such appendicitis or atypical locations (peritoneum, lungs, bladder) can occur.

Parasitological diagnosis

- Anal Scotch test: used to detect eggs in the anal margin using Graham adhesive cellophane or Scotch test technique. It is placed on a slide and examined under the microscope.

Prophylaxis

- Good food and personal hygiene (nail clipping).
- Environmental sanitation through the construction of latrines, septic tanks and water networks potable.

□ Phylum Nematoda□ Class Adenophora
□ Order Trichocephalida
□ Family Trichiuridae
□ Genus Trichuris
□ Species: Trichuris trichiura (the only species specific to humans).

Trichocephalosis or trichiosis is an intestinal helminthiasis caused by a nematode called Trichuris trichiura or Trichocephalus. It is a parasitosis linked to faecal peril and the use of human fertiliser.

2. Morphology

It is a pinkish or reddish white worm whose body is divided into two parts (Fig.39) :
□ a very thin, thread-like cephalic part, like a strand of hair, containing part of the digestive tract: a short muscular oesophagus followed by a longer glandular part made up of cells called stichosomes and occupying three quarters of the worm;
□ a much larger posterior part containing the second part of the digestive tract and the reproductive system, which in the male consists of a thread-like testis and a seminal vesicle ending in a copulatory spicule inside a sheath.

Figure 39: Adult male and female Trichuris trichiura.

Females have a thread-like ovary, an oviduct and a uterus. The vulval orifice is ventral and anterior. The male measures 30 to 35 mm with a spiral posterior. The female measures 35 to 55mm, and the posterior part is slightly arched, about 0.5mm wide.

3. The egg

It is 50µm long and 10 to 15µm wide. It is oval with a thick brown shell showing 2 protruding, pale mucous plugs at the two poles, giving it a lemon-like appearance (Fig. 40). The egg contains only one central cell when it is released into the environment and is therefore not embryonated.

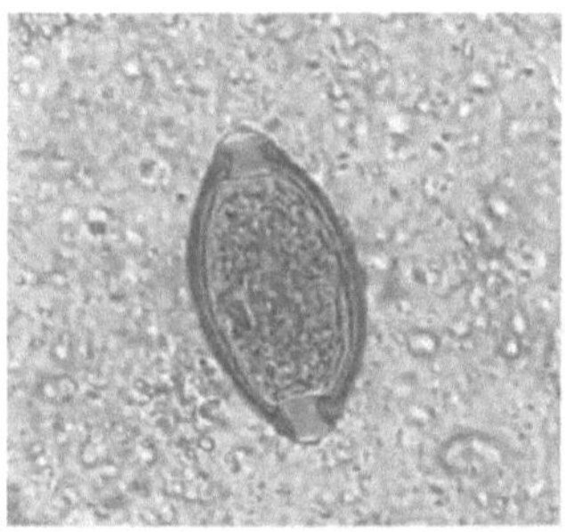

Figure 40: Trichuris trichiura egg.

B. Final host

Man is the definitive host of the Trichuris trichiura parasite.

C. Mode of contamination

Contamination occurs orally.

E. Life cycle

Eggs eliminated in faeces must remain in the outside environment to embryonate and become infective. Once embryonated, the infective eggs can survive for several years. Humans become infected by ingesting embryonated eggs infectants (Fig.41). In the duodenum, eggs hatch and release a larva. The lives in the cecum until it reaches adulthood. Whipworms live for 5 to 10 years or more, hence the accumulation of worms in susceptible individuals. The female begins to lay eggs around 1 month after infestation.

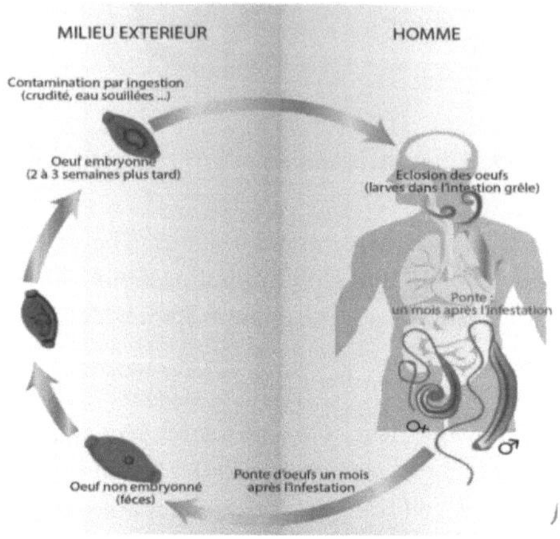

Figure 41: Evolutionary cycle of Trichuris trichiura

Clinic

The invasion : The initial invasion is rarely massive. It therefore goes unnoticed. At most, there is a hyper eosinophilia in the blood, which develops in accordance with the Lavier curve. **intestinal establishment phase**clinical expression depends on the parasite load. This makes it possible to distinguish between Trichocephalosis infection and Trichocephalosis disease:

- Trichocephalosis infection: This is the most common asymptomatic form. It is caused by the presence of 1 to 10 worms in the colon.

- Trichocephalosis disease: This is an uncommon disease with non-specific clinical signs:
- altered general condition ;
- dysenteric syndrome, nausea and flatulence, often associated with constipation or, more rarely, diarrhoea;
- behavioural problems such as nervousness and aggression;
- General signs with anaemic syndrome (iron deficiency anaemia);
- In children, there may be a delay in growth and development. Exceptionally, there may be complications in the form of :
- profuse rectal haemorrhage ;
- trichocephalic appendicitis simulating an appendicitis attack.

Biological diagnosis

The CBC showed moderate hypereosinophilia and anaemia.

Parasitological diagnosis

Parasitological examination of faeces: detection of adults or eggs.

Prophylaxis

- Combating faecal peril: use of watertight latrines.
- Treatment of human fertiliser before use.
- Drinking water supply for local authorities.
- Personal and food hygiene (especially hand hygiene).
- Screening and treatment of carriers.

□ Phylum Nematoda□ Class Secernentea□ Order Strongylida
□ Super family Ancylostomatoidae□ Family Ancylostomidae
□ Subfamily Ancylostomidae:
Genus Ancylostoma
- Species: Ancylostoma duodenale

Hookworm disease is a parasitosis caused by the presence in the human small intestine of a haematophagous nematode that is strictly human: Ancylostoma duodenale. Hookworm disease is the 4th most common parasitic infection in the world, after malaria, bilharzia and ascariasis. It occurs tropical and intertropical areas and is transmitted through faecal peril.

2. Morphology

a. Adult

These are small cylindrical worms, whitish or reddish-brown in appearance. The female measures 10 to 18 mm by 0.6 mm. The male measures 8 to 11mm by 0.5mm. The anterior end is slightly tapered with a strongyloid oesophagus (a single bulge). Ancylostoma duodenale has a mouth capsule with two pairs of hooks. There are also 2 excretory ducts and two pairs of glands that secrete a haemolytic and anticoagulant liquid. The posterior end of the male Ancylostoma duodenale worm is enlarged. It forms a large copulatory bursa, supported by 7 pairs of ribs that act as a support during mating. The posterior end of the female worm ends in a small point. The female does not have a point at the end of its tail. The female reproductive system comprises ovaries, a uterus and a vagina.In the male, there is a testis, a seminal vesicle, an ejaculatory duct and 2 copulatory spicules.

Figure 42: Ankylostoma adults

b.Eggs

The eggs measure 60µm long by 40µm wide and are clear with a thin, smooth, transparent shell. The egg is segmented into small cells called blastomeres. There is a clear, refractive space between the inside of the shell and the blastomeres.
The Ancylostoma duodenale egg contains 4 blastomeres. When the conditions are favourable (temperature 25-30°C, high humidity), the egg in the external environment will give rise to the rhabditoid larva.

c.Larvae

The rhabditoid larva is 250-300µm long, and its oesophagus has 2 bulges. Only the strongyloid larvae of the two species can be distinguished on the basis of their morphological characteristics. They have a sheath, are 600µm long with an oesophagus equal to ¼the length of the body. They have slow movements. These characteristics distinguish them from the larvae of S. stercoralis.

3. Habitat and Nutrition

Ancylostoma duodenale worms live in the duodenum. Their mouth capsules attach them to the villi of the duodenum and jejunum. Hookworms feed mainly on epithelial debris and do not assimilate blood well, Ancylostoma duodenale hookworms can live for an average of 4 to 5 years.

C. Methods of contamination

This is the cutaneous route via the encysted strongyloid larva.

E. Life cycle

Hookworms are oviparous. The eggs are eliminated in the faeces and are not embryonated when laid. In the outdoor environment, when conditions are favourable (temperature between 25 and 30°C, good humidity, sufficient oxygen, shady soil), the eggs embryonate and within 24 hours give rise to L1 rhabditoid larvae (two swellings). They feed on bacteria and organic soil debris.Within 24 hours, these motile larvae undergo a moult and transform into L2 strongyloid larvae and, 5 to 10 days later, into encysted strongyloid larvae, the form that infests humans. They can survive 2 to 6 months in the soil and more than 18 months in water without feeding. In humans, the encysted strongyloides larva becomes active and passes through the integuments via the transcutaneous route. It then penetrates the subcutaneous cellular tissues via the bloodstream or lymphatic system. This penetration generally takes place in the plantar inter-digital spaces.

It is transported to the right heart, the lungs and then the bronchioles. At this point, it transforms into an L4 larva between the 5th and 7th day. It travels up the trachea to the aerodigestive tract. During swallowing, it falls into the oesophagus, reaches the stomach and the duodenum. After a final moult, the larva grows, becomes an adult and differentiates into male or female. Fertilisation occurs and the fertilised females begin to lay eggs 40 days after infection. In the case of Ancylostoma duodenale, phenomenon of hypobiosis is described: its development is arrested and the larvae remain at this stage in a slower stage of life for around 40 weeks. They will continue to develop at a later stage. This allows them to escape to an unfavourable environment.

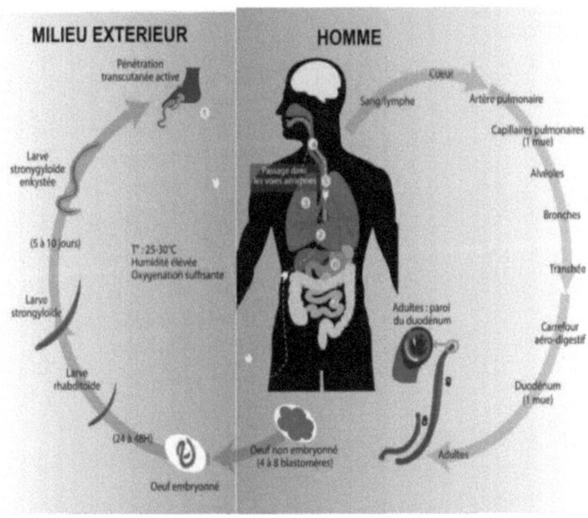

Figure 43: Life cycle of Ancylostoma duodenale

Clinic

 the larval migration phase :

- skin manifestations concomitant with larva penetration (pruritus, maculopapular erythema),
- catarrh of the gullets due to irritation of the aerodigestive tract (coughing, dysphagia, sialorrhoea). In the state or digestive phase, there are digestive disorders:
- Duodenitis, nausea, vomiting, anorexia, diarrhoea with liquid stools, sometimes mucusy but never bloody;
- An anaemic syndrome: the degree depends on the intensity and duration of the infestation

(gradual installation).

- Delayed height and weight gain children.

Biological diagnosis

FNS: iron-deficiency anaemia, sometimes discreet, with blood hypereosinophilia of 30-40% marking the larval migration phase.

C. Parasitological diagnosis Parasitological examination of stools :

Parasitological examination of the faeces allows eggs to be easily found and identified at low microscopic magnification. Eggs appear 40 days after the start of infestation.

Prophylaxis

- Construction and use of latrines
- Stopping the use of man-made fertilisers, evacuating waste water and draining land. stagnant water ;
- Avoid walking barefoot.

BIBLIOGRAPHY

ANOFEL. Parasitoses and mycoses of temperate and tropical regions. Association Française des Enseignants de Parasitologie; 3rd edition. October 2013

Brooke MM, Wilcox DE, Kaiser RL et al. Investigation of factors associated with the decline of intestinal protozoa in a Kansas mental institution. Am J Hyg 1962;76:52-6.

DELUOL Anne-Marie. Atlas de Parasitologie. Guide pratique du diagnostic microscopique. Tome II. Flagellates, infusoria and coccidia. Editions Varia, 3, rue de l'Abbe- Grégoire 75006 Paris.

MARKELL EK, VOGE M, JOHN DT. Medical Parasitology. W. B. Saunders Company. 7th Edition, 1992.

WERY M. Medical protozoology. De Boeck & Larcier. S.A., 1995.

Claude Moulinier: parasitologie et mycologie médicales, éléments de morphologie et de biologie. Editions Médicales Internationales, France 2002.

Marc Gentilini: Médecine Tropicale, published by Lavoisier, 6th edition, 2012.

Parasitoses and Mycoses of Tropical and Temperate Regions 2nd edition. ANOFEL. Edition Masson

Ripert C. Praziquantel and Fasciola hepatica infection. Tram. Roy. Soc. Trop. Med. Hyg. 1990. 84 :610-611

Sirol J, Kerfelec J, Papinutto JP. Pulmonary paragonimosis in Africa. A propos de 26 observations colligées au Cameroon occidental. I. Parasitological, anatomopathological, epidemiological and clinical facts. Bull Soc Pathol Exot, 1967, 60: 533.

M.S. Mas-Coma, J.G. Esteban & D. Bargues. Epidemiology of human fascioliasis: a review and proposed new classification. Bulletin of the World Health Organization, 1999, 77 (4): 340 346.

Morphologic descriptions of Taenia asiatica. The Korean Journal of Parasitology (1993). **31** (1): 1-6. PMID 8512894. doi:10.3347/kjp.1993.31.1.1.

Vetagro sup campus veterinaire de lyon year 2011 - Thesis n° Epidemiological study diphyllobothriosis, a parasitic zoonosis, in Haute-Savoie thesis presented at the University Claude-Bernard - Lyon I (Medicine - Pharmacy) and publicly defended on 2 December 2011 to obtain the degree of Doctor of Veterinary Medicine (Docteur Vétérinaire)

Barbara Wicht, Tetsuya Yanagida, Tomáš Scholz, Akira Ito, Juan A. Jiménez, and Jan Brabec. Multiplex PCR for Differential Identification of Broad Tapeworms (Cestoda: Diphyllobothrium) Infecting Humansv Obiamiwe BA. The pattern of parasitic infection in human gut at the.

Belen Otero-Abad, Paul R. Torgerson. A Systematic Review of the Epidemiology of Echinococcosis in Domestic and Wild Animals. PLOS Neglected Tropical Diseases PLoS Negl Trop Dis 7(6): e2249. doi:10.1371/journal.pntd.0002249

Larrieu E, Mujica G, Gauci CG, Vizcaychipi K, Seleiman M, Herrero E, et al. (2015) Pilot Field Trial of the EG95 Vaccine Against Ovine Cystic Echinococcosis in Rio Negro, Argentina: Second Study of Impact. PLoS Negl Trop Dis 9(10): e0004134. doi:10.1371/ journal.pntd.0004134.

Gentilini M, Caumes E, Danis M, Lenoble D.R, Bégué P, Touze J.E, Kerouédan D : Médecine tropicale 6éme édition, Médecine Sciences Publications, 2012 Lavoisier, Paris. ISBN 978-2-257-20396-0.

TABLE OF CONTENTS

MIX
Papier aus verantwortungsvollen Quellen
Paper from responsible sources
FSC® C105338

Printed by Books on Demand GmbH, Norderstedt / Germany